Capons for Profit

Plain Instructions To The Beginner How To Make Capons and How To Manage Them

by T. Greiner

with an introduction by Jackson Chambers

This work contains material that was originally published in 1903.

This publication is within the Public Domain.

*This edition is reprinted for educational purposes
and in accordance with all applicable Federal Laws.*

Introduction Copyright 2018 by Jackson Chambers

IMPORTANT NOTE & DISCLAIMER

IMPORTANT NOTE :

As with all reprinted books of this age that are intended to perfectly reproduce the original edition, considerable pains and effort had to be undertaken to correct fading and sometimes outright damage to existing proofs of this title.

At times, this task can be quite monumental, requiring an almost total rebuilding of some pages from digital proofs of multiple copies. Despite this, imperfections still sometimes exist in the final proof and may detract slightly from the visual appearance of the text.

Some images may suffer from reduced quality due to anomalies in the original scan.

DISCLAIMER :

Due to the age of this book, some methods or practices may have been deemed unsafe or unacceptable in the interim years. In utilizing the information herein, you do so at your own risk.

We republish antiquarian books with no judgment or revisionism, solely for their historical and cultural importance, and for educational purposes.

Self Reliance Books

Get more historic titles on animal and stock breeding, gardening and old fashioned skills by visiting us at:

http://selfreliancebooks.blogspot.com/

introduction

Here at **Self-Reliance Books** we are dedicated to bringing you the best in *dusty-old-book-knowledge* – this time, an old book on caponization.

This special edition of **Capons for Profit : Plain Instructions to the Beginner How to Make Capons and How to Manage Them** was written by T. Greiner, and first published in 1903, making it well over one-hundred years old. It is also known as **Cypher Series Book No. 6, Practical Poultry Keeping.**

The book features sections on *What a Capon is and What He Is Good For, Something About the Capon Maker, Best Birds and Breeds for the Beginner, Tools and Other Requisites, Feeding for Market,* and more.

A short, fast read and an essential addition to the libraries of all for all those interested in the historical aspect of the Poultry industry.

~ Jackson Chambers

State of Jefferson, April 2018

Kellerstrass Farm
Arthur Oscar Schilling 1907

BLACK LANGSHANS

PUBLISHER'S NOTE

THIS is Book No. 6, of the Cyphers Series on "Practical Poultry Keeping," and is an entirely new and revised edition of Capons for Profit, being re-written and brought down to date especially for this series. It is essentially a book for the beginner, and, though from the pen of an expert in the art of caponizing, it is treated from a beginner's viewpoint. "Capons for Profit" is, we believe, the only text-book on Capons and Caponizing treating the whole subject so that this profitable field of practical poultry work becomes as plain and easy to learn as the alphabet. We take pleasure in presenting this new and highly creditable addition to the Cyphers Library, feeling assured that our readers will welcome it.

Book No. 1 of the series is "Profitable Poultry Keeping in All Branches, a practical general treatise on the whole poultry industry.

Book No. 2, is "Profitable Care and Management of Poultry," telling what to do and how to do it.

Book No. 3, "Profitable Poultry Houses and Appliances," contains plans for sixty poultry buildings and many poultry plant appliances.

Book No. 4, "Profitable Egg Farming," tells all about this money making branch of poultry keeping.

Book No. 5, "Profitable Market Poultry," tells all about squab broilers, broilers, roasters, ducks, turkeys and geese.

The books of this series contain the best and most complete information on the subjects treated that has been published to date, and we believe that progressive poultrymen will find them the essential helps to success.

CYPHERS INCUBATOR COMPANY.

INTRODUCTORY.

WHEN city folks move out into the suburbs or into "the country," about the first thing they inquire about is whether they will have room enough to keep some fowls. They can get along without a cow, but they must have poultry. Why this longing for fowls? Because, aside from every consideration of profit, there is something naturally and peculiarly charming about poultry which few people who live in a rural district are able to resist. It is one of the great pleasures of country life to have a flock of nice fowls on the premises. It is a pleasure to feed and water them, and otherwise care for them. It is the delight of the women folks to sit and watch the broody biddies, or to care for the incubator, and nurse the little chicks during their babyhood, and what a lot of fun, daily renewed, it is for the youngsters to gather up the daily yield of eggs, and hunt for hidden treasures in "stolen nests"!

Who among those of us who have been "through the mill" would dare to doubt that hens, thus kept in moderate numbers, are by far the most profitable of our domestic animals, aside from the pleasures they yield, and the convenience of having superior food materials, in fresh eggs and poultry meat, always on hand. The home poultry surely pays well—always has paid well, and certainly always will pay.

With many people, however, it has been a question whether poultry kept in large numbers will pay or not. The great factors in the case which determine the outcome more than any other, are management and feeding.

One of my neighbors, a young man who owns a few acres of land, perhaps eight or nine, has for ten or a dozen years made a better living, and much more easily, than the farmers around him who own or work moderate sized farms and make general farming or in part trucking their business. How does he do it? Merely and solely by keeping a respectable flock of hens for egg production. He has managed to do this even under theoretically unfavorable conditions and surroundings. His hens numbering between 400 and 600, were year in and year out kept together in one flock, a reasonably large barn being arranged for their accommodation. It

5

was a mongrel lot. Leghorn blood, however, predominating. With neghbors' houses on both sides within five or six rods, it has always been necessary to keep the flock more or less confined to an orchard lot surrounded by a high picket fence.

This man always got the eggs, and although they were sold to regular customers at only a slight advance over current market prices, they paid him well for his trouble and expense.

He had learned three things. One was to feed a properly balanced ration, compounded from materials that happened to be offered "on the bargain counter" I might say. Another was to keep his flocks and the building reasonably free from lice and mites, and the third was to have no fetid matter accumulating and no stagnant pools of water standing on the premises where fowls might get to them to contract filth diseases. There can be no Godliness without cleanliness, and surely no success in poultry keeping, and no profits. Cleanliness, which, of course, includes a continuous supply of fresh water, and freedom from vermin, must be named as the very first requisite. The next is the proper food supply.

In the hatching business, my neighbor has thus far followed the old-fashioned plan of

setting hens, giving them a separate room for the proper performance of their task. Two broods of chicks were usually put together with one hen in a coop outdoors. The young cockerels are always killed off for sale during the summer months, while the older hens are selected for slaughter as soon as they cease laying in mid-summer or early fall, to make room for the pullets.

The business method of the present day, of course, is to hatch the chicks in good incubators, and bring them up in good brooders. Both contrivances have now been brought to a degree of perfection which eliminates the chances of failure, if fresh and fertile eggs are used, and the machines and chicks managed with ordinary judgment and care. I shall not say much about this part of the business, as it can be safely left to the manufacturers of the best incubators and brooders to tell their own story.

Neither am I going to boom any breeder of thorough-bred fowls, or any particular breed of fowls. I must leave that also to the breeders themselves. Every good business man, every one who has a really good article to offer, must know how to advertise his wares, and it is decidedly proper that he should do so to the best of his ability. My neighbor operates with mongrel fowls, and as he is successful with them it would be unwise in the extreme to advise him to change to pure bred fowls, or to any particular breed. "Let well enough alone," may be a good motto in his case.

As for me, I want a uniform flock, all of one color and of one build. A flock of mongrels, fowls of all sizes, all shapes, all colors and shades, may show variety, but it is not a particularly interesting or pleasing sight. It seems to be an aimless, purposeless congregation of things. Then go and see a flock of Black Langshans, or somebody else's flock of

White Leghorns, or Wyandottes, or Barred Plymouth Rocks, or Pekin Ducks, or White Holland Turkeys, or Bremen Geese—or even White swans! What an imposing sight! How attractive! How beautiful! How pleasing!

And don't imagine that the pure bred stock is less profitable than the mongrel. Far from it. Our beautiful thoroughbreds lay just as many eggs as the mongrel stock, and often, perhaps usually, more. They have plumper bodies, and they are generally more valuable.

Nearly twenty years ago I accidentally drifted into the Black Langshans; they seemed to suit my particular purposes and notions better than any of the other breeds, of which I have tried a large number. They have their faults as have others. Their skin is white and not yellow. They make the poorest kind of spring chicken, being slow to "fill out," and often very unattractive when dressed, all on account of their dark pin feathers. But they suit me for their docility, their egg-laying qualities, their superiority as mothers, their hardiness, size and fine table qualities. The imposing beauty of the flock when at its best appeals to every visitor or passerby. During all these years I have never lacked the chance to sell nearly every egg we had to spare during the breeding season at double or treble the market price for ordinary eggs.

And if I had cared to raise breeders for sale, they would have brought twice or more what they are worth for meat purposes. Why should I keep mongrels? Why should anybody?

"Capons for Profit" is the title which I have selected for this book. It suggests that the chief purpose we have in view in making capons is that of financial gain. In this connection two points of view may be taken. One is from the standpoint of the home poultry keeper, the other from that of the specialist who raises capons on a more extensive scale for market. I am prepared to state that capons are profitable in either case, but before going more deeply into the merits of this question, some general explanation in regard to capons and the capon industry will be required.

T. GREINER.

CHAPTER I.

WAHT A CAPON IS AND WHAT HE IS GOOD FOR.

THE only justification which a poultry keeper has, or could give, for keeping and feeding a male bird, popularly known as "old rooster," is the fact that his services are needed for purposes of propagation. The aim of the fowl owner who keeps poultry for practical purposes and not merely for ornament is to get eggs and raise chicks. Every male bird kept in excess of the necessary requirement for the continuation of his race is a damage to his owner, a useless eater, and without one redeeming quality.

The number of hens that may be safely kept with one male bird in order to insure fertility of the eggs and strong chicks is an oft disputed point. For years I have had only one rooster to from thirty to fifty hens, and always had fertile eggs and strong chicks; so I have made up my mind that this proportion of the male to the female element is sufficient, at least for the Langshan breed. If one rooster will do why should I waste food and care on two or three or more, and let them fight over the affection of their flocks to their mutual damage besides? Yet this is exactly what most poultry keepers and poultry owners are doing. The country is full of useless roosters kept in excess of requirements. Where hens are kept to lay eggs for table use, not for hatching, and especially for long keeping, or laying down for winter, we had better go a step further, and not only reduce the relative number of males but dispense with them altogether; the hens will lay just as well, or even better, and the eggs will keep much longer in prime condition.

The chief object I have in view as a "Home-Poultry Keeper" is to produce a great abundance of eggs for a large family—all, in fact, that could possibly be wanted summer and winter, and all the poultry meat that we can make use of. We do not easily tire of eggs, nor of really prime poultry meat, so long as these things are prepared by a skillful cook, in a variety of ways, in one way one time and in another way another time.

To accomplish our purpose fully, i. e., to have pullets ready to lay in the fall and all winter, and also to have the required number of cockerels for meat stock, we are obliged to raise 100 to 150 chicks a year, beginning early in the season, say March, to set eggs under hens or in an incubator. If the latter, we have two great advantages, viz., we get through the hatching business within a few weeks' time, and don't have to bother with a lot of sitting hens during a period of two or three months, and we have early birds which will make early "spring" chicks, early layers, and very large capons.

One half the number of chicks in our flocks are usually males. The Langshan "spring" chicken is nothing to brag over. It could hardly be expected to bring highest price in market, on account of its poor and "pin-feathery" appearance. The pullets are pushed right along to make early layers.

The cockerels, except the very few that may be needed for breeders, and which have to be selected for their general appearance and promise while yet small, are made into capons. By keeping a lot of cockerels over until nearly full grown, I might have a chance to sell a proportion of them at $1.00 to $3.00 each for breeding purposes. In most cases I would have to sell one or two pullets with each male, and it is not my plan to raise early pullets for sale. They are too valuable as winter layers, unless I could sell them at a fancy figure. For that trade I would have to keep a lot of roosters until spring,—birds which if not wanted for breeding are about as useless and valueless as any creature in fowldom.

The surplus roosters, in short, are a nuisance on the place, always mischievous, harassing the laying hens and reducing the profits, and at best they will bring only six or eight cents a pound, that has cost you more than that amount to produce. My way now is to turn them into capons, and thus treble the value of their flesh.

There are a great many persons in America who have never heard of capons, and have not the least idea what they are.

"What a magnificent lot of fowls!"

"Say, Mister, what breed are your fowls?"

"Will you sell me a trio of your fowls, or a setting of eggs in the spring?"

These are the questions that I and some of my friends who keep a goodly number of capons have been asked time and time again; and we had to explain the matter as best we could. If we simply said "They are capons," people would answer, "Why, yes; they are just the breed we want." I think myself that they will want them after they once find out how tender and sweet and juicy their flesh is.

But what are capons, anyhow? Farmers make oxen of their surplus male calves, and wethers of their surplus male lambs, and ordinary horses of their male colts. In short, they castrate or emasculate all male animals not wanted for breeding purposes. Occasionally male cats and dogs are treated in the same way; and the process of castration gives us better cats and dogs than they are in their unaltered state. We used to castrate male rabbits when we were breeding them for table use, and it improved them greatly in size and quality of flesh, and as in other animals, made them more peaceable and quiet.

But while thus improving by castration the surplus males of all farm stock, farmers have neglected the male chickens, probably considering them beneath their notice in this respect.

This is a big mistake.

It is easier to castrate a young cockerel than a pig or a lamb.

It is profitable in more ways than one.

Once operated on, capons become the most tractable and peaceable animals imaginable. They do not run, nor chase, nor fight. All they seem to live for is to eat and grow and become fat. I will not say that capons while young grow faster, or lay on more flesh from a given amount of food than unaltered males of the same age do. As long as the cockerel is young enough so that no energy goes into the reproductive organs, I think their development is about at an even rate. But there is a change after a time. The development of the organs of reproduction in the male, and his growing activity and restlessness consume energy, which is saved in the capon for flesh production.

Water only comes to a certain degree of heat. All the excess above this is utilized for the formation of steam. It is the same thing with the rooster. He grows to a certain size or weight, and all the surplus energy above this is used for the purpose of reproduction. In other words, the capon will continue to grow and lay on flesh much longer than the unaltered male. It takes a year or more for the capon to come near his full size and weight, but even at that age he is usually heavier than the rooster of the same age, and several times more valuable. A pound of old rooster is worth from six to eight cents; a pound of capon from 18 to 28 cents.

I raise capons chiefly for home consumption, as already suggested. In fact, without my supply of capons, I would miss a good share of my table enjoyments, and the pleasures of other rural privileges.

We may use a few springers during June and July, and a few of the old hens (that have to give way to pullets by another spring) during August and perhaps September, after their heavy work in egg laying is done. But during eight months of each year my capons furnish the meat material for my roast chicken, chicken potpies and chicken soups. An eight or ten pound Langshan capon is not a bad substitute for the roast turkey on the Thanksgiving dinner table, or for the "baked goose" on Christmas.

On Sundays or festive occasions we may entertain company at dinner. A large capon furnishes plenty of the choicest meat food for six or eight persons. When the family is alone, or only one or two friends take dinner with us, we have the capon roasted with dressing, and with the proper relishes in celery and cabbage salad, cranberry or currant sauce, etc., and the meat that is left over, with the abundance of rich gravy, is served for our Monday dinner with fresh biscuits, in same manner as rural people often serve "chicken with biscuits." Both meals are eminently enjoyable and enjoyed.

I have never figured out what amount in cash my capons save me in the course of a year that would otherwise be spent for butcher's meat. I know that it is a large amount, roast beef costing us from 12 to 18 cents a pound and good steak even more. Surely there is profit in capons for home use.

I have also had some experience in growing capons for the city market, while a brother of mine has marketed moderate flocks in a Western New York sanitarium, where the demand was much larger than he could supply, with prices ranging fom 18 to 22 cents and upwards per pound. The question is what are the prospective profits in an undertaking of growing capons in large numbers for the open market.

I might ask what are the prospective profits in raising turkeys in large numbers? The profits may be entirely missing, and again they may be considerable. It depends on management, judicious buying and use of foods, etc.

Capons have material advantages over turkeys. The chicks are easily produced or procured. If we fail to hatch the chicks we may go out over the back roads in July, and soon pick up a load of Brahma. Langshan, Cochin or Plymouth Rock chicks, or chicks of

10

other large breeds, at twenty or twenty-five cents a piece, and of just the right size and age for caponizing. The operation is a simple matter, and with us or anybody using reasonable care, uniformly successful. The young capons are great eaters, with an appetite for anything that comes along, a most peaceable disposition and rugged constitution. They are seldom attacked by disease. They grow to nearly the size of a turkey, and when ready for market bring more money, pound for pound. It does not take an extra large capon to bring $1.50, and $2.00 is often realized for one bird.

The capon has one disadvantage. It must be carried through at least a part of and perhaps the entire winter so that it may reach its full size and bring the largest price. It requires housing like hens, but it can stand a good deal of crowding.

Some capon makers claim that not more than forty or fifty birds should be kept in one flock. We have had nearly one hundred of them together in rather crowded quarters, and they did well. In a general way, however, I would advise that they be given roomy quarters, free from draughts, reasonably protected and well ventilated, and otherwise managed in same manner as skillful poultry keepers manage hens for laying.

A Group of Capons.

Chapter II.

SOMETHING ABOUT THE CAPON MAKER—THE MAN IN THE CASE— WHO IS FIT TO OPERATE AND WHO IS NOT.

THERE are people whom I would not advise to undertake the operation of caponizing. The person to do it should have, above all things, faith in his undertaking and in himself. He must be convinced that his work is right, and then go ahead. This is no place to make a trial for fun, or in a half-hearted way. It is a little of the genuine enthusiasm that is needed, and that is bound to overcome difficulties should any be encountered. Then there should be an average amount of mechanical skill and the same amount of nerve. Clumsy fingers have no business operating on a live fowl. If you are a little nervous at first it will do no hurt. Your nervousness will wear off after you operate on two or three fowls, and see how easy the job is, and how little pain it apparently causes to the bird. You must have full confidence in your ability to do it just right, and then go ahead without fear or trembling. And when you are once at it, it is far better to operate on all the fowls ready for the operation in one day than fuss along with two or three every few days. The beginner is apt to be a little nervous when he goes for the first bird; but after he gets his hand in once, everything moves off smoothly and nicely. His hand becomes steady and the work passes off rapidly. Of course, it is an advantage if you can see some one perform the operation, even on a single bird. No more is needed to teach you the whole operation. But the average person does not often have a chance to see it done.

I had to learn it from books and printed instructions—not very plain ones either—and succeeded beyond my expectations. After a few days' practical experience you will think lightly of the achievement of caponizing 20 birds a forenoon. The operation, indeed, after you have once undertaken it and succeeded, is an easy enough thing, and causes but little pain and inconvenience to the bird, if you do it right and with proper tools. A good set of tools, of course, is utterly indispensable, and the person unwilling to expend $2.00 or $3.00 for them is not included in the list of persons who can safely undertake the operation.

Then there are some timid souls who tremblingly ask, "Does it hurt?" The farmer when castrating lambs, or pigs, or calves, etc., does not ask, "Is it cruel?" The butcher, when he kills a sheep, or hog, or calf, does not stop inquiringly, "Does it hurt?" Both know well enough that they do inflict some pain to their victims, but they yield to the demands of necessity. We can not always avoid suffering pain, or giving pain.

What is cruelty? The needless infliction of pain. This is cruelty and decidedly a wrong. Yet many persons, too tender-hearted to stick a hog, or to castrate a pig, or to have a boil on their own back lanced, will, when provoked, use the whip freely on their children, or kick and strike their horses and cows most unmercifully on slightest provocation.

All this is a needless infliction of pain, and therefore cruelty and inhumanity. We can and should be merciful and decent, all the more when we are compelled to make other creatures suffer. It is not necessary to insert a hog-hook into a hog's mouth and pull the animal into the scalding vat before it is dead. It is not necessary to begin skinning a calf or lamb when yet alive. I think these things are horrible, and people of any heart and feeling would not stoop to do things so mean.

In short, the person whom I would like to induce to turn mischievous, worthless, cockerels into peaceable and valuable capons, is the one who is impressed with the necessity and advantages of the operation; who has faith in his abilities, an ordinary amount of mechanical dexterity and nerve, a little energy and perseverance, and is in possession of the tools needed for the operation. This

man is neither a sensitive weakling nor a cruel brute.

People too morbidly sensitive to use the knife on a live bird; people with clumsy fingers, people without proper tools, or people who are brutes, are the people whom I would wish to keep their hands off this business.

To reassure and encourage those timid people, however, who are afraid of undertaking the operation simply because they imagine that it inflicts severe and cruel pain on the bird, I wish to state that in reality it causes discomfort, for a few minutes, rather than actual suffering. In the summer when one of the poultry editors came to see me operate on some of my cockerels, to convince him that the injury to the bird is not excessively painful, I gave him ocular demonstration of the fact that the pangs of hunger make the bird forget the pain caused by any part of the operation. Some birds will pick

A Typical White Wyandotte.

up food placed within their easy reach on the table, even while a testicle is being removed. When the poultry editor left he was fully satisfied that chicks are not very sensitive to pain of this kind.

The man to succeed, if he cannot learn the operation by seeing somebody else perform it, must be guided by the right sort of directions. In our younger days, for many years we (brother and myself) talked about, and wished to learn, the operation of caponizing, and tried it time and time again, mostly on dead chicks. But we had the books of instruction written by advocates of the method of removing both testicles from one side. We invariably failed, and I believe that nine out of every ten persons who will open a bird only on one side and try to remove both testicles out of the one opening, will fail. Even now, with all the practical experience I have had, I would surely give up the job in despair if I were obliged to cut from one side only. There is a big difference in the interior construction of chicks even of the same breed, the same strain, the same

age. Sometimes both testicles will appear in plain view from one side, in which case I invariably remove them from that side. In other cases you will find it almost impossible to get a view of the lower testicle, no matter how much you push the bowels back and aside. Instead of torturing the bird for a prolonged period by probing in its inside and fishing for the lower testicle, I simply turn the bird over, make a second incision, find and remove the testicle, all in a few minutes' time.

I cannot make this advice too strong. Whatever you do, as a learner and beginner, begin right. Open both sides of the bird, and you will succeed. Don't listen to the people who tell you of the added cruelties of the second incision and manipulation; it will lead you into trouble and failure. The cruelty is a fake, or merely imaginary. When we got hold of the Dow method of cutting birds on both sides, our troubles were over and we succeeded at once; adopt the same method and your success will be assured.

Buff Plymouth Rocks.

14

Light Brahmas.

Chapter III.

THE VICTIMS AND THE TABLES—BEST BIRDS AND BEST BREEDS FOR THE BEGINNER—SIMPLE OPERATING TABLES.

THERE is a great difference in breeds and birds. For many years I have taken an especial fancy to the Langshan breed, and the cockerels with which I had to make my first trials were either the pure Black Langshan or crosses of Langshan cocks on Plymouth Rock hens. Of all breeds I have tried I find the Langshan the easiest subject to operate on, because the bird makes bone first and flesh afterward, in other words, is usually lean when young, showing the ribs quite prominently. It offers little difficulty to the prompt removal of the testicles, and apparently is suffering the least while under the operation. Besides this, the Langshan has the advantage of large size and great hardiness. It is also less liable to make what is known as "slips" than most other breeds.

My next choice would be the Langshan and Plymouth Rock cross. The cockerels, in plumage and outward appearance, resemble Plymouth Rocks quite closely, yet

15

offer about as little difficulty to the novice as the pure Langshan. They make large, noble-looking capons. Most of the ordinary mixed fowls of our barn-yards are easily operated on. Cochins I have never tried. Of course, they are large and will make good capons. White Wyandottes have given us great satisfaction as capons, too.

Brahmas will grow to largest size, and

Barred Plymouth Rocks.

may prove the most profitable of all breeds for this purpose, yet the beginner will be apt to have trouble with them. The ribs do not show prominently on the outside. Although this makes little difference to a person after he has operated on a number of fowls, it may puzzle the beginner. The most serious stumbling block, however, is the shape of the testicle, which in young Brahma cockerels is about a half inch long, extending close and worm-like along the big artery. My emphatic advice, therefore, is to make the first trial with easy caponizers, especially the Langshan or Langshan cross, or with ordinary smaller breeds, never with Brahmas. [Many expert producers of capons do not agree with Mr. Greiner on this point. In the vicinity of Hingham, Mass., where the famous South Shore chickens are produced for special fancy trade in the Boston market, the Light Brahma is by far the most popular breed. The male birds are invariably caponized for the production of fancy soft roasting chickens, and from the testimony of these men, who are making thousands of Brahma capons yearly, we are led to believe that Brahmas are easily made into first-class capons if taken at the proper stage of development. We have been repeatedly assured that it is no more difficult to caponize a Brahma cockerel than one of any other breed. In fact, many professional operators consider the Light Brahma one of the easiest to caponize. —EDITOR.]

The following is taken from a bulletin of the New York State Experiment station at Geneva:

"It is better to use only the larger breeds for capons, and the Brahmas and Cochins are among the best, but while these breeds furnish poultry of superior size and excellent quality there is compared to the game an undesirable deficiency of breast development which is plainly noticeable in the dressed fowl. At the New York Poultry Show in 1892, the first prize was given by a competent judge to a capon eight and one-half months old of Indian Game–Buff Cochin cross over capons young and old of Light Brahma, Black Langshan and two or three other breeds and crosses. A cross of the Indian Game gives nearly as large fowls as the pure breed with much of the game shape. This cross can probably be used with advantage, for the Indian Game while larger than the Pit Game has little of the fighting spirit of the latter and having yellow skin and legs will not interfere with the common prejudice in that direction. It is not probable, however, that did such prejudice exist in a market demanding the best of capons, it would be hard to overcome where good fowls of such breeds as the Dorking, Houdan, La Fleche and Langshan were to be had."

The Indian Games excel in quality, and will probably make the choicest capons for home use. The markets, however, do not yet discriminate between capons of different breeds except so far as size is concerned. For profit we want the fowls that will give us the greatest weight–Langshan, Cochin, Brahma, Plymouth Rock.

You may be sure that whenever I have a chance to buy a coop of young Brahma or Cochin cockerels during the caponizing season at a reasonable price, I take them quickly, and convert them into profitable capon stock.

I also find that it is less trouble to operate on comparatively young subjects than on older and larger ones. When I want an easy job I take a two pound Langshan, Langshan cross or Plymouth Rock. My Langshan-Plymouth Rock crosses seldom flinched even when the incisions were made or the testicles twisted off, while Brahmas, when taken at

White Plymouth Rocks.

more advanced age and size offer more or less resistance, and must be held more firmly.

The best weight, for Brahmas, to operate on was usually given as four pounds or more. I cannot see the reason why we should let Brahmas get that big. At any rate I operate on them while not much over two pounds in weight, and usually succeed with them as well as with birds of other breeds.

It is also a good plan to use a dead subject for the first lesson. Shut the victim up without food or drink for 36 hours. This is important, as you want the intestines empty. Then chop his head off, put him on the operating table in good light, and otherwise in the same way as will be described for the operation on a live subject, and go ahead making your observations in cockerel anatomy.

An empty barrel, bottom side up, may be

Indian Games.

made to answer for a table, the fowl being held by means of one stout twine tied around the wings next the body and another tied around the legs, the free ends of both hanging down on the side of the barrel and weighted with a brick or piece of iron. I would put padding of some kind, a piece of old carpet or a bag, upon the barrel head under the chick, thus giving him as comfortable a rest as possible under the circumstances. This kind of operating-table, however, is a poor make-shift at best.

When you have a large number of cockerels to operate on, or set out to caponize your surplus roosters for profit, year after year, as you should, you will want a more convenient table. Dow, and others, advise you to have a table made for this special purpose in as simple a style as you please, with cleats around the top at the right to prevent the tools from falling off, a two inch hole in the center at the left, with a weighted lever underneath and a mortise six to eight inches long from right to left, also in about the center of the table, with a sliding lever, weighted underneath. A twine loop is fastened on each one of the levers, passed up through auger-hole or mortise, and slipped one over the wings, the other over the feet, thus securely holding the subject for operation.

You can also make a table such as shown in Fig. 1. It consists of a round board larger than a barrel-head, resting on an empty, headless barrel. Weighted straps or bands are drawn through two holes bored at proper distances, and hold the chicks, wings and legs, as may be seen in the picture. This table has the advantage that you can turn it toward the light to suit, without moving the barrel. But it affords no good chance to place the tools and is not particularly handy.

The table I use is illustrated in Fig. 2. It is a light, cheap kitchen table, such as we happened to have to spare, three and a half feet long and twenty-two inches wide, more than large enough to accommodate the cockerel and leave plenty of room for the tools and yet light enough to be easily shifted about for the sake of getting the light just right upon the work. I fastened some

Fig. 1. The Barrel as an Operating Table.

narrow cleats with screws all along the margin of the right-handed half of the table, thus rendering this part a safe place for the tools and accessories. At the middle of the opposite (short) side, screwed into the edge, is a screw-eye or hook, which holds the loop of twine after the latter is slipped around the wings of the victim next to its body. Its legs are held by a strip of board, which is padded with flannel on the under side, and weighted on top with a piece of iron or a brick securely fastened with wire or twine. One end of this lever is cut in convenient shape for a handle, while the other has a cleat which simply hooks over a longer cleat screwed fast upon the table. This arrangement allows the lever to be moved sideways, according to the size of the fowl, or entirely taken off when the table is not in use. The cleats may also be removed by taking out the screws, and the table be put back where it belongs, in kitchen, buttery or cellar.

I always place a piece of old carpet, an old fertilizer sack or something similar under the fowl, doubled or rolled up to extra thicknesses under the legs, thereby securing a close fit and a firm hold without unnecessary pressure upon the fowl's legs between hard objects.

All the details of arranging an operating table we may safely leave to the good judgment and the common sense of the operator himself. I do not claim that my table is the only one that could be recommended, or the best for its purpose. The whole matter is extremely simple. The purpose of an operating table is two fold, viz.: First, to provide a convenient place where the bird may be held firmly in the desired position, without undue crowding or cramping, and shifted table and all, to get best position for light; and, second, to have a chance to place the operating implements and fixings where always handy and accessible. Where a great deal of caponizing is to be done, I would arrange an open case with a place for each kind of tool, where it may be held by a wire spring or in some such way, so that the operator can put his hands on any tool he may want without having to hunt the table over for it.

The table as I use it is selected for convenient size and lightness, and as above described, it can be made ready for business in the simplest manner, and in a few minutes' time. If you prefer to use different ways and details, there is no objection. Hold the bird firmly on the table, and that is about all that is necessary for success.

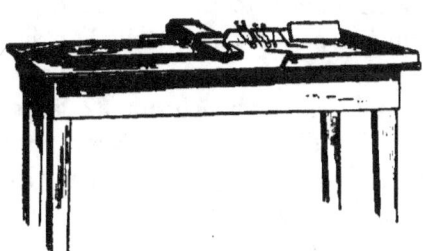

Fig. 2. Greiner's Operating Table.

White Wyandottes.

CHAPTER IV.

TOOLS AND OTHER REQUISITES—WHAT IMPLEMENTS ARE BEST SUITED FOR THE BEGINNER.

LET the motto be "Good tools for good work." In an emergency an expert can get along with and do fairly good work with an inferior set of tools and equipments. The beginner cannot. It is not safe for him to undertake to do a delicate operation with clumsy tools. A set of caponizing instruments need not necessarily be expensive. Indeed such tools as I have found most serviceable, can be bought for $2.75 per set. I have tried different makers' sets. Some of them you can use, and do good work with, and others are entirely unsuited to the beginner's needs and requirements. For that reason I have settled on a combination known as the "Greiner set," which I will here describe. First is the lance or knife. In an emer-

gency an ordinary pocket knife, sharpened to a razor edge, might answer, but it has not the best shape for the work, and is a makeshift at best. A lance is shown in Fig. 3, and made of a piece of steel one-sixteenth of an inch thick, seven sixteenths of an inch wide and about six inches long. Have an oil stone handy and keep the lance well sharpened. The knife as made by some manufacturers has a straight edge, as shown in Fig. 4. I prefer to have it well rounded, as then you can make the incision with one dip, and yet without having to go very deep with the point as you have to do with the straight-edged knife.

Next you need a spreader. A whale-bone spreader was formerly much used. We now

20

Fig. 3. The Knife or Lance.

Fig. 7. Nippers and Forceps.

have wire spring spreaders that are much simpler, and much more convenient to use. In fact, I think the simpler the spreader is, the better. The one shown in Fig. 5, gives good satisfaction. It should have a maximum spread at the end, of at least three quarters of an inch. Still simpler and cheaper

Fig. 4. Straight Edge Knife.

is the wire spread shown in Fig. 6. There are various other styles of spreaders that can be safely used, but those here illustrated are as serviceable as any, and have the advantage of simplicity.

I prefer the former style, which seems to be far more convenient and less in the way of the

Fig. 5. The Best Spreader.

operator's hand and tools. Its extreme length is fully seven inches. The other which is only about three inches long, often interferes with good work.

Next comes a pair of nippers, or forceps, something like Fig. 7. It is used to hold a

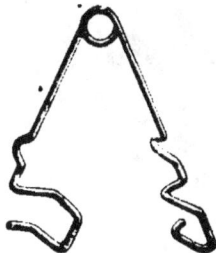

Fig. 6. Spreader.

little piece of sponge with which to mop up blood inside the capon, or to pick up any stray object that may have fallen on or among the intestines. Almost any ordinary small nippers will answer the purpose.

The set should also contain a small, sharp steel hook (Fig. 8) and a probe (Fig. 9). The former is used to tear open the thin, film-like membrane which envelopes the intestines; the latter to push the intestines back when crowding over the testicle, or in the search for any object fallen among the intestines.

Figs. 8-9. Hook and Probe.

The most important of all caponizing implements, however, is the canula, with which to catch and remove the testicles. Spoon nippers or forceps are often used in place of a canula, but they are not safe in the hands of a beginner, and I would not use them under any circumstances. To operate on small (two-pound) cockerels of breeds that have not the worm-like testicles of the Brahma, I prefer a small, straight canula operated with fine, flexible wire.

This is a brass, nickel or aluminum tube, about $4\frac{1}{4}$ inches long, a quarter of an inch wide at the large, open end and tapering to

The Canula.

21

Buff Cochin Cock.

about three thirty-seconds of an inch. It is closed at the small end, with the exception of two holes large enough to admit a small steel wire.

Years ago we used horsehair (from the tail) altogether, and in an emergency I would and could now use it with good success. I find fine, flexible wire, such as bee keepers use to fasten comb foundations in their brood frames far superior to horsehair, and the best by far of anything that I know of for such purpose. It is liable to kink. But so is horsehair. It is stiffer than horsehair, and we have less trouble to get it adjusted around the organ that we wish to remove. This wire is very cheap, and and a five-cent spool of it will last pretty near a lifetime. After it has become badly kinked, we simply throw it away and insert another piece. The little wire coils usually sent with the caponizing sets on the market are not what I would use or recommend. The wire is too heavy and stiff. It wants very fine, flexible wire.

The birds we operate on are usually small. If we use a large, clumsy canula, we take up so much room with it inside the small body and organs, that we can't see very well what we are doing. It is for this reason especially, that I want and recommend a small canula. The testicles are frequently not larger than a kernel of wheat, or a small soy bean. Even a small wire loop can be slipped around it with ease. But in case the bird is older, or we operate on Brahmas and other fowls having a long, worm-like testicle, it might be well to use one of the Pilling canulas, which is a $\frac{3}{8}$-inch tube, flattened and slightly bent at one end. It admits of having the loop wider and more convenient for catching on the long testicle.

I think it is well enough to have a canula of this kind on hand, especially if you operate on Brahmas, etc. Still, the other canula will do in an emergency.

Of other requisites you will need a sponge or piece of sponge, and a few cents' worth of carbolic acid and the wire. That is about all.

Once more let me say, don't try to get along with clumsy imitations and substitutes. For good work, and satisfactory work, you need good tools. If you wish to caponize at all—may the number of cockerels to be operated on be half a dozen or a thousand—the first thing

Buff Cochin Hen.

22

to do is to get a complete set of tools. They are cheap enough, and they will last you a life time.

If you once learn the operation, easy as it is, you will have calls from neighbors and others, and possibly you may find a little work at better wages than is paid for ordinary farm work. After you have learned to caponize, I am sure you will not allow a worthless lot of roosters to infest your premises, bother your laying hens, and eat three times more than they will be worth in the end.

If you do not feel able to invest the small amount for tools, don't undertake the job of caponizing. For the sake of humanity and decency, don't murder helpless animals with clumsy tools. Be merciful. When properly performed, and with the tools here described and illustrated, the operation involves no element of cruelty. The birds seem to suffer slightly when the incision is made, but for a moment only, and again when the testicle is twisted off, but at no other time, and they are ready to take their meals as soon as the job is done.

For those operators who have a great deal of caponizing to do, and may desire to use even dark days, or hours during the twenty-four when the sun is near the horizon, the use of artificial light was suggested by the Rhode Island State Experiment Station. This is a comparatively simple matter. You may use any ordinary physician's head mirror, that can be secured at any dealer's of surgical instruments. It is not even necessary to use electric light, or acetylene light, in connection with it. For $2.00, you can buy a physician's head mirror. Have a lamp, covered by means of a paper hood so that the flame shows only through a round hole cut in paper hood, placed conveniently so that the rays reflected from the mirror carried just above the forehead, may be focused into the opening made in the bird's side, and all the organs inside appear fully exposed to plain view.

As for myself, I do not feel the necessity of using artificial light on ordinary summer days, even if somewhat cloudy.

The novice, of course, will need all the best light he can get, and nothing is much better than a subdued direct sunlight on a clear day, especially when the sun is midway between the horizon and the zenith, or say at from 9 to 11 in the forenoon, and from 2 to 4 in the afternoon in midsummer. I or any other person, after some years of experience, can do work just as easily on a rather dark day, or in the early morning and late afternoon hours, as at any other time. Yet at times, one may find "more light" desirable in treating particular cases or particular birds, and I believe one of the electric flash lights now much used by policemen, would come very handy.

A Sectional Brooder.

A Group of Light Brahma Capons.

Chapter V.

THE OPERATION—WHEN AND HOW BEST TO PERFORM IT.

THE three months for capon making are July, August and September, but the operation may be performed at any time when you have the right material for it.

The first, and an absolutely necessary thing to do, is to catch the cockerels to be operated on, and shut them up in a convenient coop, so that they can easily be gotten hold of when wanted. Do this in the evening, and then leave them for about 36 hours without food or water. The intestines should be fairly empty in order to enable the operator to use the knife without fear of cutting into them, and to give a good chance for work and for seeing what one is about. The long fast will not hurt the chicks, but only makes them very anxious for the next meal.

On the morning of the second day, when the sun is two or three hours high, and the sky nearly or entirely cloudless, the operation may begin. The expert can manage to get along without much direct sunlight, but the light cannot be too good for the beginner. A clear day is absolutely necessary for a first trial, and if the day should be dark, the cockerels may be given a *very small quantity* of soft food, to carry them over **to** the next (supposedly clear) day.

Set the caponizing **ta**ble in a convenient spot and in direct sunlight, or on a clear, hot day of midsummer, perhaps under the rather open branches of some solitary tree, the foliage of which permits the passage of occasional rays of sun, giving a slightly subdued or modified but direct sunlight.

24

In all these things, of course, good judgment should be consulted.

Spread the tools out on the cleat-enclosed part of the table. On another table, stand, barrel or box, close by, have a dish with warm water medicated with a few drops of carbolic acid, also a large piece and half a dozen small pieces of sponge. The latter may be of about the size of robins' eggs or hickorynuts.

Now pick out the first victim. Let it be a rather lean bird, weighing not more than two pounds, nor much less. Twist or wind the twine loop around the wings close to the body, and standing in front of the table, with the cleat-enclosed end to the right, fasten your victim on his left side upon the table, as shown in Fig. 10. Next to the hip, and where, in a lean bird, the ribs show quite

Fig. 10. The Bird Ready for Operating.

plainly, you find a spot which, because, usually covered by the wings, is almost bare. There may be a dozen or two of pin feathers. These should be pulled out. Take hold of them between thumb and index finger, not one by one, but as many as you can take, and deftly pull them out. Don't be nervous. Go at it as if you meant business. If you are quick and determined about it, the removal of these small feathers does not cause much inconvenience to the bird, for the latter never makes any fuss over it. The spot thus cleared need not be more than one and one-half inches in diameter.

If a few short pin feathers remain that cannot easily be caught up between the thumb and finger, use the nippers to remove them,

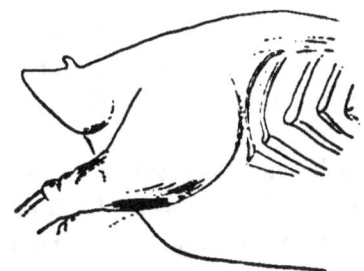

Fig. 11. Diagram of Ribs.

or use the point of thumb and the knife to catch them up and remove them.

At this stage of the proceedings I take the sponge out of the water, squeeze most of the water out of it, and then wipe it over the chicken's side. This is not absolutely necessary, but it moistens the feathers around the bare spot, and helps to keep them out of the way.

Now comes the incision. The right place to cut is between the last two ribs; that is, the two ribs next to the hip. In a lean chicken they are easily recognized, and often they are very prominent. They extend from the back bone for an inch and a half or two inches in a slight curve, then take a sudden turn upward toward the breast. Usually the "joints" in the two ribs appear plainly and prominently. Just look for the two slightly raised, whitish, almost knob-like spots. Often the two ribs lie quite closely together, and perhaps the end of the muscle—a flat layer of flesh—extends over them.

An examination of Fig. 11 will give you a pretty clear idea where to look for the spot. Here the location of ribs is plainly shown. The dotted line between the last two ribs is the right place for making the incision.

Fig. 12. How to Hold the Knife.

Now proceed by taking the knife in the right hand as shown in Fig. 12. Then with the left hand, reaching over the right, push the skin and muscle from the bared spot toward the hip and hold it there. Observe the two little whitish spots which form the joints of the last two ribs, and set the point of the knife right between them, making an incision by a quick dip, at the same time slightly drawing the knife between the two ribs toward the back bone. The length of the incision should be about one inch. With the intestines nearly empty there is no danger of injury to them, even if the point of the knife should dip half an inch deep through the ribs. Minor blood-vessels usually extend in the skin across where the incision is to be made. If they are cut, a few drops of blood will be spilled; that is all. But in pushing skin and muscle toward the hip, and drawing it tightly, you may at the same time aim to get the blood-vessels somewhat out of the way of the knife. If this is done, the knife often does not draw a drop of blood. If the wound bleeds badly, the moistened sponge may be pressed upon it for an instant to absorb the blood. Making the incision, of course, will cause a momentary pain to the bird, but it is of no more than any living thing has to endure a good many times in life, and will do so without complaining.

After the incision is made, lay down the

knife and take up the spreader, all the while holding the skin back toward the hip with the left hand. Press the spring of the spreader together until the two free ends meet, and then insert them in the opening and let go. Also release the skin yet held with the left hand. The spreader will push the ribs apart, leaving an opening to the fowl's inside from one-half to three-quarters of an inch wide. If the cut was not large enough, you can remedy it by a slight touch of the knife to one or both ends of the incision.

From now on in the proceedings you will need good light. Shift the table about, or turn it as required, so that the best light will reach into the opening and upon your work. Looking down through the incision, you will notice a thin, translucent film or membrane, which covers the entire internal organs. The little blood which may have dropped in from the outside wound and clotted on this membrane, is most easily removed by picking up, with the tweezers or forceps, a little piece of moistened sponge, introducing it into the opening, and pulling it out again with all the blood adhering to it. The membrane now appears clean and translucent. Then take up the steel-hook and carefully pick this membrane to pieces, always holding the sharp point of the hook upward, or in the direction of the back-bone, in order to avoid touching the organs that may be crowding against it from below. The tear through the membrane must be large enough to expose, under good light, the internal organs to view. When bowels are nearly empty, you will plainly see, well toward the back-bone, the upper testicle, a yellowish body of about pea size (of course, larger in older cockerels), perhaps somewhat elongated, or in the Brahma, etc., quite long, almost worm-like. Sometimes both testicles come in plain view, especially if you push the intestines aside with the probe or with a similar tool. Sometimes, again, it happens that the intestines crowd upon the upper testicle and hide it from view. Then introduce the probe and push them aside, and the testicle will come in full view. Its light color (although often it is partially dark-colored, almost black) makes it plainly visible.

You are now coming to the object of all this proceeding; namely, the removal of the testicle. Take up the canula. The wire should previously have been adjusted to form a loop of nearly three-eights of an inch in diameter. Slip this loop over the testicle, and between it and the big artery which may be seen alongside of testicle. If at first you don't succeed, try again. It may require several trials, but don't lose patience. It will go all right at last, and usually the testicle will be caught up easily at the very first trial. When you see that the loop has properly caught on, draw up on the loose ends of the wire, at the upper end of canula, so that the loop is all pulled in, and the testicle tightly drawn up to the end of canula. Hold the canula with the left hand, twisting it back and forth about half way around, at the same time pulling continuously and strongly on the ends of the wire with the right hand, thus cutting and twisting the testicle off its fastenings. When you feel it give way, pull it up with the canula and wire, and if some of the strings still adhere to it when you get it up through the opening, cut them off with the knife so that a little bit, say 1-32 of an inch, remains on the testicle. This is important. If you cut too close to the testicle, nature may try to thwart your purpose by letting a new growth of testicle take place, thus causing what is known as a "slip."

The thing to be avoided is injury to the big artery. If the blood-vessel should form a kink, and the kink be drawn into the horsehair loop, the artery will be torn, and the fowl will bleed to death in a few minutes. With reasonable care, however, this does not often happen.

I have not lost one per cent. of the birds that I have operated on during the last five years. Usually the testicle in small birds becomes detached from its fastenings quite easily, and can be pulled out of the opening, picked up and thrown away. Occasionally it may get loose, and drop back into the interior of the bird, in which case you simply take the nippers, pick the organ up from among the intestines, and remove it. If it gets mixed up with the intestines, you have to probe and "fish" for it until you find it and can remove it. Should you notice blood gathering up inside the bird, as is occasionally the case even where no important blood vessel is ruptured, introduce a little piece of moist sponge, using the probe to push it down among the intestines, and then pull it out by means of the nippers; squeeze out the blood, or take another piece of moist sponge and repeat, until the blood is mopped up and the flow has ceased.

Now one side is done. All that remains to be done is to see that no feathers or other foreign substance are left inside the opening; then take out the spreader, let the skin and muscle slip back over the incision through the ribs, unfasten the chick, and— turn him around on the other side for another operation.

I have described the job in all its minutest details. To perform the operation does not require one-tenth the time that it takes to tell

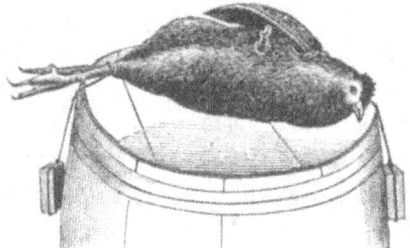

The Wire Spreader in Place after the Incision.

it. In the first attempt you may possibly spend a quarter of an hour or more. What does it matter? Take your time. The fowl, while not especially comfortable, is not actually suffering. He feels slight actual pain only during the moment when the incision is made, and perhaps during the removal of the testicle. After you have operated on two or three birds the task becomes an easy one and and the operation will not take many minutes. The difficulty is only in the first attempt. Expert operators usually remove both testicles from the one opening, the lower one first and afterwards the upper one. It is not safe for the novice to attempt it.

The turning of the bird is quickly done. Lift up the lever, taking hold of the chick's legs, turn him over on his right side, as shown

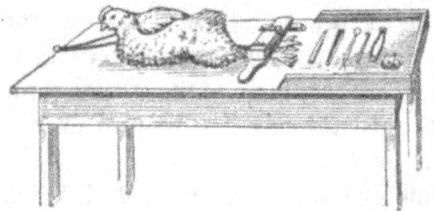

Fig. 13. The Bird Ready for the Second Operation.

in Fig. 13, and readjust the lever to hold his feet. Again shift the table so the light will fall fully upon the front of the fowl, and into the opening to be made on the left side. The operator this time stands on the other side of the table, next to the chick's back, as before. Then a few feathers are plucked out, the moist sponge wiped over the bared spot and the surrounding plumage, the incision made and the whole operation gone through with in exactly the same manner as was done on the other side. The fowl is placed in a rather more convenient place to be operated than when lying on its left side. A good deal of bleeding is sometimes going on after the testicle is removed. While a little blood if left inside among the bowels, would probably do

The Cyphers Marker.

no harm, there may be more than the system can absorb in a natural way, and the clotted gore might harden and cause inflammation, blood-poisoning and death. At any rate it is advisable to remove this blood by the means already mentioned; namely, mopping up with little bits of moistened sponge. Sometimes you will have to introduce one bit after another, half a dozen or even a dozen times, letting the sponge soak up, then withdrawing it with the tweezers, and squeezing it out in warm water. At other times there may not be a drop of blood spilled. If that is the case, or otherwise when the flow of blood ceases, see that no feather, bit of sponge or other foreign article is left inside; then withdraw the spreader and unloosen the bird. Next mark it in some way to show that it has undergone the operation. One of the easiest and quickest ways to do that is to clip off the end of one toe. All my capons have a stub inside toe on the left foot. The style of marking is a matter of individual choice. You can use one of the 25-cent or 50-cent chicken markers, two styles of which are here shown with which to punch a hole or two through the web between two toes; or you might use leg bands, or any other of the various devices. A stub toe, however, suits my purpose as well as anything. Put the fowl's foot upon the table, hold a keen knife blade across end of toe to be amputated, with knife point down upon the table, and then with a quick move press down the handle of the knife, like a lever and thus

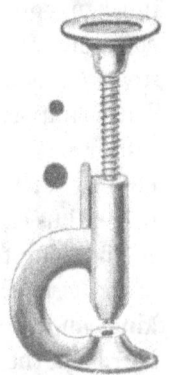

The Philadelphia Marker.

clip off the toe end. It seems to make no particular difference to the fowl, either at the time that it is done or afterwards.

Of course, you will occasionally lose a bird in your first trial. When in removing the testicle you rupture a big blood vessel, the bird will die under your hands, usually in less than ten minutes. In that case, chop his head off, if you so prefer, and have a fine spring-chicken dinner. Accidents of this kind, however, should not occur oftener than once in fifty operations on ordinary fowls, even with the beginner. But if they do occur there is no loss, as the bird has its full value for table use. A capon that comes out alive after the operation, is "out of the woods," and complete recovery is swift.

Capons in Colony House at Farrar Bros., West Norwell, Mass.

White Wyandottes "Bred for Business."

CHAPTER VI.

THE AFTER TREATMENT—HOW TO HASTEN THE HEALING PROCESS.

MY elaborate description of the whole operation of caponizing may lead many of my readers to imagine that the capon is in bad shape when he comes from the operating table. This is a mistake. The testicles are not a vital part. Their removal is of little consequence so far as the bird's health and vitality are concerned the only difference it makes is in regard to the sexual development of the bird. The thin membrane which envelops the intestines is not a vital part. The holes which we have torn in it are an injury which amounts to almost nothing. The only real injuries inflicted, therefore, are the two incisions, and these are merely flesh wounds of the simplest character, and by no means severe. The skin, having slipped back in its natural position, covers the flesh wound between the ribs; the cut in the skin, and that in the flesh, on each side, cannot well be in more favorable shape for rapid healing. There is no need of sewing the edges of the wound together, or using any kind of salve, or plaster, or wash. Just trust

in Nature, the best physician, and you will not be disappointed.

At the beginning of the caponizing season, I used to put up what I called my "capon hospital." This is simply a space on one side of the barn containing, say, two or three square feet of floor surface to each capon, and covered with a low roof as a protection from rain and sun, and tightly enclosed with wire netting. Inside is a coop in which the convalescent fowls spend the nights. A box in one corner is kept well supplied with soft food (bran and meal moisted with skim-milk), and a dish in the other corner contains the water which should be frequently renewed. Some of our instructors tell us to feed lightly at first; others advise giving all the food that the birds will eat. I usually have tried to keep the box supplied with food all the time, but it is a hard task, for the birds have a keen appetite and eat a great deal. Just as soon as a capon is put into the "hospital," and once gets sight of the feed-box, he will forget all the trials which he has just experienced, and at once proceed to fill his crop. The birds may be confined in close quarters for at least eight days after the operation. Many of them "wind-puff" badly, a lot of air gathering under the outside skin and giving the capon a puffed-up appearance and probably causing much inconvenience. I usually looked the confined birds over once a day, and gave speedy relief where needed by pricking the puffed-up skin with the point of a keen penknife. Part of the birds do not need this attention; others wind-puff right along for a week or so, and need frequent touches with the pen-knife. Usually you can tell by the appearance, and always by the feeling, whether there is wind-puff and cause for treatment. In consequence of this confinement, of the treatment and of their voracious appetite, the capons become exceedingly tame and tractable.

My later practice is to give to the young bird his full liberty right after the operation, letting him run, feed and roost with the rest of the fowls. Neither this liberty, nor the dry (grain) feed, nor want of prompt attention, when wind-puffed, seem to retard his perfect convalescence. If I note an especially puffed-up appearance, I try to catch the bird and give him relief with the knife. Still, I be-

lieve it is a good plan to keep the birds confined for from eight to ten days, giving soft food and proper attention otherwise. The straw, leaves or soft earth on the floor of the "hospital" should, for the sake of cleanliness, be often renewed.

When the period of convalescence (eight or ten days) are past, the capons should have their liberty. They will not wander off very far, but stay most of the time near where they are accustomed to get their regular rations. At night drive them into the "capon house," a warm stall with low roosts, regularly cleaned and disinfected. Capons do not seem to be particular about their roosting place. I try to keep my hens and capons apart at night, hen house and capon house being only separated by the width of the barn. But when a capon happens to be nearer the hen roost at the time he wants to retire for the night, he forgets where his place is, and unhesitatingly takes lodgings with the hens.

How many capons may be crowded together in one building seems to be a disputed point. Some experts claim that 40 to 50 is the limit of safety. We have had 80 to 100 roosting together in a rather small house, the capons having free range in day time, with satisfactory results.

Profitable Capons.

Capons in Colony Houses, Jordan Farm, Higham, Mass.

CHAPTER VII.

FEEDING FOR MARKET—HOW TO OBTAIN BEST RESULTS AT LEAST COST.

THE chief aim, from the time the bird is caponized to the time of sale or slaughter, should be to produce the heaviest possible weight, and for this reason a liberal supply of flesh-forming food should be given. During the summer and earlier fall months I feed mostly bran mixed with corn meal and slightly moistened with skim-milk or butter-milk, and whole wheat. Corn is not a proper food then; but some variation is provided for by giving an occasional mess of peas, buckwheat or oats. My fowls have free range, and find good pasture on the lawn and in a piece of rye and rape sown for this very purpose close to the barn. Grasshoppers, bugs, worms, table-scraps, etc., all help to fill the fowls' crops and to produce capon meat. A vessel in the yard is kept supplied with skim-milk almost all the time.

All quick-growing animals have good appetites, and young capons seem to be always hungry. Notwithstanding their tendency to laziness, they are good foragers.

The problem of profitable feeding during the summer and early fall, indeed during all mild and open weather, is comparatively an easy one to solve. Fowls on free range find so much to pick up in nice warm weather that small additional rations of grain will suffice to keep them in good growing condition. The natural advantages seem to me all in favor of that climate which allows fowls to be out on pasture the greatest number of days during the year; and if I were to make it my chief business to raise "capons for profit," I think I would try to locate in a country with mild, dry, open winters, and on dry, sandy soil. With us in the North, the problem of feeding capons grows in degree of complication and difficulty with the severity of the winter. When we aim for largest size in capons, as we should, we will have to keep them until they

32

are about one year old. Usually there is little demand in our markets for capons until midwinter or early spring. Before that time they would not bring much higher prices than ordinary fowls. After that time the prices range from 18 cents per pound upwards. In short, if we wish to secure largest possible size of fowl, and highest possible price per pound, we have to keep the capons all winter and perhaps until spring. Now anybody who has ever wintered fowls on purchased food, knows that they eat a great deal, and that the bills for grain, even when wheat is only 60 or 65 cents, and corn 50 cents a bushel, soon run up to a large amount.

The trouble is that many people think grain is the only, or even chief, poultry food. This is an error. Exclusive grain diet is not only expensive, but also unnatural and unsafe. It may do well for a week, when fowls are being fattened for slaughter, but if long continued, it will surely clog the system, make fowls over-fat, and in the end injure their general health and well-being. It should be understood that the tendency in capons, especially in cold and stormy weather when kept in enforced idleness, is to grow and lay on fat. Even without excessive feeding they are bound to get fat as butter.

What is needed, in the first place, is a cheap, bulky material that will fill the fowls' crops, taking the place of grass and leaves of summer. Then we want a little moderate amount of grain to add substance and warmth in place of the weed seeds and the like found scattered about in the open season, and finally, something in place of bugs and worms.

The bulky material is best and most cheaply supplied in chopped vegetables and chopped clover hay. Every fall I store a lot of beets, carrots, turnips, kohlrabi, etc., in the cellar, and cabbages in the barn or out-doors, for the very purpose of utilizing these vegetables for winter poultry food. Cabbages are for the most part simply thrown into the hen coops or capon houses as needed. Roots of all kinds, also small potatoes and apples, are chopped up in a plank box with a sharp spade, or in one of the regular root cutters of poultry size, or in a vegetable chopper, then (sometimes slightly salted) mixed with a little bran and fed in their raw state. In cold weather a mess of beets, turnips, carrots, pumpkins, squashes, small potatoes and peelings of all kinds cooked in a big kettle, and stirred up with bran to a thick, crumbly mass are greatly relished by all fowls as a warm breakfast or dinner. Chopped clover hay, and the chopped leaves of corn stalks, are scalded, sprinkled with bran, and then fed warm. In short, with materials of this kind we can keep the birds' crops well filled at a very small cost.

But one can make this food even much richer at little additional expense. Green bones, with more or less meat on them, are a waste product of butcher shops. The proprietors usually are glad if somebody comes to take it away. At any rate this richest and (when fed in reasonable limits) best of all poultry foods can be had at a very slight expense. It is a most excellent substitute for the bugs and worms of summer. It is a pity so much of it is wasted.

The question only arises how we can get the hard bones, and the tough gristle, and other fleshy matter fine enough for fowls to eat. The often-advertised $5.00 bone mills will not grind green bones. Two ways are open for us to utilize these waste products. They can be softened by steaming under high pressure. It might pay us to get a steam cooker suited for the purpose. On the whole, however, I believe that it is preferable to cut bones, gristle, flesh, etc., with one of the cutters now especially designed for that purpose. The accompanying illustration shows one of these machines. It does not *grind*, but will cut or shave any of the materials named in pieces fine enough for fowls to eat. Where many hundreds have to be fed, a power machine will be preferable.

Now and then a little grain—wheat, oats, buckwheat, etc., should be given, and the evening meal should always consist of whole grain, chiefly of corn in very cold weather. If plenty of the cheaper and more bulky food is given, four quarts of whole grain would be enough for one hundred fowls.

The habit of eating the feathers off around the neck sometimes develops in capons as well as in other fowls. This may be due to idleness in confined birds. It will occasionally, though rarely, occur in flocks that are given free range. Keep the birds busy in search of insects on pasture, etc., and there will be very little trouble from this source.

"The feather eating habit," says the New York Experiment Station, (Geneva, N. Y.), in an earlier annual report, "is more often, perhaps, induced by improper food, lack of

A Root Cutter.

animal food, or lack of variety in the ration. Laying hens fed at this station about two months almost exclusively upon Indian corn and corn meal, picked not only the feathers but flesh from each other, so that two were killed. This same trouble has been seen elsewhere when birds were closely confined, with little chance or inducement for exercise and no change in food. A pen of young capons and "slips" was fed for about two months exclusively on corn and corn meal (plenty of fresh water being at all times available), and these birds picked nearly all the smaller feathers from each other. An entirely similar pen, fed at the same time a mixed grain ration, continued in perfect plumage.

The birds afflicted with this bad habit were easily cured at the Station by the application of lard or vaseline (sometimes one and sometimes the other) in which had been mixed powdered aloes, to the old feathers near the spots which had been picked bare, and on the new feathers which appeared. After continuing this treatment for some time, the habit apparently disappeared so that the birds were enabled to grow a full coat of new feathers. No change of any consequence was made in the food, etc., and the suppression of the habit was probably due to the disagreeable taste of the aloes.

It is thought that an extract of aloes would probably be better where grease on the feathers is objectionable."

Some feeding experiments with capons have been recorded by the New York State Experiment Station at Geneva. The report says:

"The much higher prices at which capons are quoted compared with those of the average of poultry, have led to many inquiries being made during the past few years in regard to the profit in growing them for the market. When we remember that beef cattle have been fed in this State during recent years at very small profit and that often to find any profit in producing pork it has been necessary to take into account the advantage of using skim-milk, etc., and to consider the manurial value of the grain fed, we may find it well worth while to learn the cost of any possible animal product of the farm that will command a good price in the market."

The trials spoken of were made with several lots of capons for the months during which they are usually grown, beginning in August and September when young cockerels are old enough for caponizing, and continuing until February, at which time the birds are so nearly mature and the growth becomes so slow that it is only a question of holding them longer or not, for higher prices.

The average weight of the birds in the Station trials was 3.8 pounds at the time of the operation. I believe this was a serious mistake at the very start. The operation should be performed at a much earlier age, and when the bird weighs not over two pounds. At the larger size, there is much greater danger from the operation, but even after being successfully performed, the loss consequent upon the thirty-six hours' fasting and the operation is much larger. Our young birds suffer little, and lose but imperceptibly in weight, and in a week's time after the operation have already made large gains in weight. Again I quote from the Station Report:

"There is a certain per cent. of loss from various diseases, accidents, etc., among the fowls at all ages, which it is important should not be forgotten in making estimates of profit, but as this varies so much with the character of the stock and their quarters, care, etc., there can hardly be any average assumed; but it is safe to say that with favorable conditions and careful attention a loss with young

chicks or older birds of five per cent. can be expected.

"No bird among those grown at this Station has died during two years past directly on account of the operation. (The loss of one some weeks after caponizing, was due more to an oversight in after treatment than to the direct effect of the operation itself). But in order to make sure of killing no birds it is occasionally necessary to leave some with almost the assurance of their developing into slips. Even the most expert professional operators expect to kill a few birds.

"The excess that the average market prices show over the cost for food is enough to promise a fair profit, over an ordinary per cent of loss, for any reasonable investment of labor, etc. The cost of caponizing, where the services of any expert operator can be obtained is but a few cents per fowl (sometimes as low as four cents). After a fall in the high broiler prices of spring and early summer, it will probably be found more profitable to caponize the surplus cockerels than to

A Convenient Food Cooker

market them, especially where cheap skim-milk and grain is to be turned into a market product. For while often the percent of profit over the cost of food in selling at broiler age is greatest, the actual difference per fowl in market price over food is greater with the capon, providing the latter is sold before growth has ceased. After caponizing the labor in caring for and feeding is but little more than in feeding cattle or pigs, and the proportion of labor to produce 100 pounds of capons is, therefore, less than in the production of 100 pounds of broilers as the latter have most of the time been with the hen or brooders.

"In the first season's feeding there was fed skim-milk (both Cooley and Separator) wheat, corn meal, alfalfa forage, dry bone and mixed grain, which was made of five parts wheat bran, one part N. P. linseed meal, one part wheat middlings and four parts ground oats by weight. There was fed the second season skim-milk, wheat, corn, alfalfa forage, beets, corn silage and two grain mixtures, containing, by weight, five parts corn meal and one part each of ground oats, wheat bran, wheat middlings, and N. P. linseed meal, and another which contained the same as the first,

Another Type of Food Cooker.

except that two parts of O. P. linseed meal were substituted for the one part N. P.

"The ratio of protein to non-nitrogenous constitutents in the ration was about such as is generally found to give good results, and was usually that of about 1:4 or 1:5.

"The largest breeds will be found the most profitable for capons, and it is useless to caponize cockerels of the smaller breeds.

"Skim-milk can be profitably fed to capons and if sweet, in large quantities. If sour, very little should be fed. It is very important that the dishes from which milk is fed should be cleaned often and scalded occasionally.

"A variety of food should be given to capons as well as to other fowls, and rations somewhat similar to those fed in these experiments will give good results. With equally good lots of birds, rations differing somewhat (but not excessively) in the proportion of nitrogenous to non-nitrogenous constituents will not make much difference in the growth. A ration containing some corn meal will in general be found to give better results.

"The cost of feeding capons after they have nearly reached their full size is approximately five cents per day for each 100 pounds live weight. The advisability of holding those of middle-weight breeds after reaching seven to eight pounds weight or the larger breeds after reaching nine to ten pounds weight will depend upon the prices to be obtained."

Mr. M. P. Wheeler of the New York State Station has also recorded comparative trials of the merits of whole and ground grains, and gives the following summary:

"A ration consisting mostly of the ordinary ground grain foods and containing no whole grain was more profitably fed to chicks than another ration consisting mostly of whole grain and containing no ground grain.

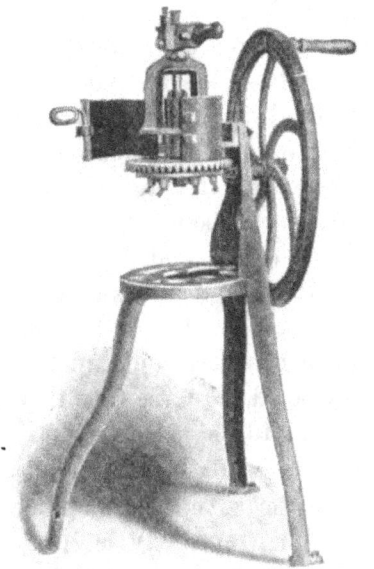

Hand Bone Cutter.

"Capons from the one lot afterward made a somewhat cheaper gain in weight on the whole grain ration, but the gain was too slow to compensate for the more rapid growth which had been made, as chicks, by the lot having the ground grain ration.

"Of two other lots of capons, those having the ground grain ration made the more profitable gain during several months.

"In every trial more food was eaten when the ground grain was fed than when the whole grain was fed.

"Neither the chicks and capons having only the whole grain nor those having only the ground grain showed any lack of health and vigor."

In feeding capons as well as laying hens, or young chicks for any purpose, I believe in making every effort to obtain cheap food materials. In some cases, table and kitchen waste can be obtained at nearby hotels. Then again, grains that are a little "off" in quality, such as shrunken wheat, or damaged grain of any kind, wheat and corn that have suffered in elevator fires, spotted peas and beans, etc.,

may often be bought at low rates, and you should be on the lookout for bargains. Meat meals, having a high percentage of protein, are now being put on the market at a comparatively low price. The richest of all foods, rich especially in the expensive protein, (the blood and muscle-forming materials), are peas and beans, including soy beans and southern cow peas. In fact, there is no vegetable food known that has a higher feeding value than soy beans. Quite often, these grains, being slightly damaged, have but little market value, and may be purchased at a mere song. Ordinary beans that may be too badly spotted for sale, are still as valuable for feeding as the A, No. 1 article. At times I have bought second grade peas, (ordinary Canada field peas) at 50 cents per bushel, and by the carload they have been offered by Western growers at 70 cents per bushel. They are worth double that amount for poultry food, and may be fed whole or

Power Bone Cutter.

ground. Fowls soon learn to like them. Still richer in protein are common beans. When a lot that is slightly damaged, can be had at less than a dollar per bushel, they become available for feeding capons or other fowls. Of course, they should either be ground, together with other grains, especially corn, and then made into cake or mush; or mash mixed with boiled potatoes. In this shape they constitute a grand concentrated food that will make plenty of meat at the lowest cost.

Still richer are cow peas (cow beans), and even above them soy beans. Both of these may be used in somewhat the same way as Canada field peas. Fowls learn to like them even in their natural condition, and I have frequently fed capons and laying hens with whole cow peas and soy beans once a day right along, four quarts to 100 fowls. It is very likely, however, that even better results may be obtained from these grains when ground and mixed with corn and oats, and fed as already suggested for feeding ground common beans.

Clover Cutter.

For green forage, as also for dry feeding, we have three leguminous crops of great value —alfalfa, crimson clover and hairy or sand vetch, alfalfa being by far the greatest among them, and for green forage alone we have Dwarf Essex rape and ordinary winter rye.— A little patch of rape will give a lot of poultry pasture during fall and early winter, while the fowls will find good picking in a rye patch in early spring, usually too late, however, to be of much benefit to our capons which are mostly disposed of by the time that the birds can get out into the fields again after winter. This, to some extent, is true of crimson clover, which, however, gives pasture late in the fall. At that time, rape may be preferable, although it can hardly be said to give a quality of meat for the true epicure. Rape has the cabbage flavor, and may impart some of it to the capon meat if the bird is killed soon after the period of rape pasture.

Crimson clover, cut early, or when just in bloom, makes a most excellent hay that can be used for cutting, steaming, etc., and mixing with meals for capon food during the winter. But for an all-round plant to serve both as green forage and hay for winter feeding, alfalfa is king, ranking high up above everything else. It gives green pasture from early spring until winter—a pasture that is unequaled by any other forage plant known for high feeding value. You can let the fowls pick the leaves themselves, or you may cut the crop and throw it into their pens and yards. It also gives a hay that is unequaled by any other hay for feed, having a percentage of protein almost equal to the best wheat bran.

It is true that alfalfa wants rich and well-drained soil, and considerable nursing and pampering at the start. But you can find a dry spot of rich ground near your poultry houses, a spot that can be enclosed to keep

Hallock Food and Water Holder (Open for Cleaning).

the fowls out for a while, and where you can sow alfalfa seed on a well-prepared seed bed, and give it proper attention for a while, cutting it early and often with a mower or scythe until the plants are well established. They will then be good for years, and give an amount of the choicest green food that will astonish you if you have never grown alfalfa before. The equal of five or six tons of wheat bran are easily grown per acre, and this without any effort, cultivation or feeding after the first year.

In reference to the advice before quoted, that sour milk should be given sparingly, I will add that my way of utilizing most of the nutrients contained in sour milk when we have a large quantity of it available for the purpose, is to make curds of it. All fowls eat it ravenously. The whey may be added to the swill for the pigs, or thrown away.

One more food material may be mentioned. It is available especially within easy reach of Niagara Falls, where there is a palatial factory of so-called "natural food products." It is here known as the "Perkey Plant," and its chief product is the "shredded wheat biscuit." The waste product consists of broken biscuits and crumbs, all broken up fine, resembling the dessicated cocoanut of our stores in shape. This waste product has about the same analysis as whole wheat, but is in even better

shape and of higher feeding value. At $1.00 per 100 pounds, it seems not only a reasonably cheap food, equaling 50 to 55 cents per bushel of wheat, but also a most excellent and wholesome one. It is of especial merit as a food for little chicks, both when moistened with milk or water or fed dry. I often use it to mix with cooked mashed potatoes or roots or to mix with other meals, animal meal, etc., milk, buttermilk, etc., for a chicken cake.

That fresh water should at all times be kept within reach of the birds need hardly be expressly stated. They should have no chance to drink out of stagnant pools in the barnyard. Sweet skim-milk, however, can be largely or almost wholly made to take the place of water. We can safely give to our capons all they will drink. One of the best devices for holding milk, water or solid food, for fowls, that I have ever used or seen, is the Hallock Patent Food and Water Holder, here illustrated. It comes in different sizes, but the largest size is just right for our flocks of capons. We want one for every twelve to fifteen fowls, if used for giving solid food; if used only as drinking fountains, one for thirty to fifty fowls will be found sufficient. If preferred, home-made devices may be used.

One of the devices of the kind that I have used with a good deal of satisfaction, for feeding the daily "mash," and which anybody of ordinary mechanical turn of mind can easily construct, is here illustrated.

Hallock Food and Water Holder, (Closed, Ready for Use).

It is made of odd pieces of boards and slats, and of any length desired. One of the roof boards is hinged so that it can be thrown back when we wish to put the feed into the trough. The slats are placed just wide enough for the fowls to stand rather close together; a space of two and a half inches between them being sufficient.

Another thing needful is a free and continuous supply of sharp grit. I do not think that there is anything superior to raw limestone broken into pieces of pea size or smaller, although ground oyster shells or sand containing coarser particles will answer.

This grit is as necessary for satisfactory results as food. Without the required grit, food cannot be properly digested, and a large part of it will be wasted.

At the edge of Niagara River, where the water is shallow and the shore sandy, we find great quantities of small shells, more or less broken and having sharp edges. We can get them by the wagon load, or by boxes or barrelsful if we get them by boat. Poultry, if given free access to this material, will consume a great deal of it, picking out all the sharp pieces. From coarsely sifted coal ashes fowls also pick out a great many of the coarser particles. Something of this sort, however, is absolutely needed, and it will usually be a good investment to purchase the Pearl grit or small stones offered at a comparatively small price by poultry supply dealers. The illustration shows a convenient device by which shells and grit can always be kept before the birds. Home-made devices, of course, may be used for the same purpose.

A similar hopper may be used for keeping dry food, especially a mixture of grains, such as wheat, barley and oats, right before the capons all the time. Or the grain may be kept in one box, or division, and beef scraps in the other. A good dry food hopper is here illustrated.

Greiner's Home-Made Feeding Device.

CHAPTER VIII.

HOW TO KILL, DRESS AND PACK CAPONS.

IN the matter of selling capons, the same general rules' should be followed that are applicable to that of selling any other farm produce.

First of all keep track of the market.

How?

By reading the market quotations; by visiting the large city markets if you live near, or by corresponding with commission merchants, wholesalers, big hotels, sanatariums and similar institutes which may be expected to have use for capons in quantity.

If feasible, have a definite understanding with some large buyer or dealer. Often there may be found chances of contracting for the delivery of a certain number of capons at stated intervals at a stated price. If not, act on advice and suggestions received from your middleman or buyer. The size of the demand depends somewhat on the size of the capon. Large birds are usually much sought for.

When you are ready for the disposal of a lot of capons; keep them in confinement at least twenty-four hours before killing, entirely withholding food and water. You want their crops to be quite empty.

You need a thin-bladed, sharppointed knife, the regular French killing knife which retails for 50 cents, being, perhaps, preferable to all others. It is here illustrated. For killing and dressing, select or arrange a cool, welllighted shed. Fasten two pieces of strong twine to a beam overhead, say a foot apart, and let the lower ends, each of which has a "slipping noose," hang down to within three or four feet from the ground, or just at proper height to be most convenient for plucking the fowl. Fasten one of the capon's legs in each of the nooses, letting his head hang downward. Then holding the knife in the right hand, take hold of the bird's head with the left, and open his bill widely. If the light shines well into the throat, you may quite plainly see the big jugular vein on each side. Insert the point of the knife, and sever these blood vessels by a quick cut across. The blood will at once follow the knife, and flow freely in two big streams. Without further delay run the point of the knife through the roof of the mouth clear into the brain. The bird now is so near dead as to be without sense of feeling. Just at this time, also, the feathers

Capons Dressed for Market.

Capons Dressed as Soft Roasters.

off the feathers as quickly as possible. Leave on the feathers of the head and neck (hackle-feathers) all the tail feathers, with a few feathers up toward the back, and the long feathers on the hips close to tail, also the feathers on the legs half way up the "drum-sticks." Be careful not to tear the skin.

This peculiar style of dressing is a kind of trade-mark, and serves to distinguish the capon at once from any other fowl in market. The head, with its shrunken or non-developed comb and wattles, should always be left on the bird. Remove all traces of blood from his head and mouth by careful washing with cold water.

A table should be on hand upon which to dress the bird, also a trough or box without ends and cover, just large enough to hold him in best position, back downward, for the removal of the intestines. Carefully cut around the vent, and pull out the intestines. You will find a heavy layer of fat around the opening

come off quite easily, and no time should be lost. Of course the bird flutters and struggles. A bird with its head cut off, and consequently without consciousness and feeling, does the same thing. In order to keep it with head down, and comparatively quiet, it has been advised to suspend a two-pound weight attached to a hook from the bird's upper bill. We dislike to see so much blood spilled on the floor, and besides being wasted; as for these reasons we hit on the device of using, in place of the two-pound weight, a two-quart tin pail containing a quantity of meal. When sufficient blood has accumulated, the pail is emptied, meal, blood and all, into a larger vessel, and again partly filled with meal. The blood and meal mixture, well stirred together, makes a most excellent food for other capons, or for laying hens.

Anyone going about this work, should of course, be clad in old clothes, or be supplied with rubber apron, and in readiness for a muss. Take good hold of the bird, and strip

Picker at Work.

and the intestines covered with it. As you pull out the intestines, push the fat back into the bird, and when you come to the end of the intestines push your finger up into the body, along the intestines, and separate them from the gizzard, leaving everything else inside the bird. [Note: For most markets capons should not be "drawn," (the intestines removed) except in very warm summer weather. During cool weather undrawn poultry keeps better and commands a better price, *Ed.*]. The layer of fat around the vent may be turned slightly outward and thus allowed to cool and harden. The bird will look all the better for it. Then hang him up in a cool place. When thoroughly cold he will be ready for boxing and marketing.

New, clean boxes should be used for shipping capons. Nothing, in fact, must be neglected to make the packages as well as the capons look neat and attractive. Line the boxes with clean, white paper. Printed paper should never be used. Pack the birds down in layers, backs up, as solidly as can be, yet without bruising. You will have no difficulty in finding a market for nice, large, fat, well-dressed and well-packed capons at acceptable and profitable prices.

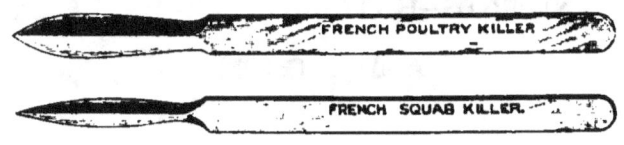

French Killing Knives.

CHAPTER IX.

SOME ODDS AND ENDS—HATCHING AND HATCHERS, BROODING AND BROODERS.

I have already touched on the matter of hatching, both by natural and artificial methods. The modern incubator has been brought to a high degree of perfection, and success in the poultry business now hinges far less than formerly on the ability of the operator to manage his incubators. This perfection of incubators solves the problem of getting satisfactory hatches for the operator. If you will look out for the right sort of eggs for hatching, the machine will pretty nearly do the rest—that is, provided you have one of the most approved style of hatchers, like the Cyphers, that can be depended upon to do good work. If you wish to hatch eggs by artificial means, by all means get the very best incubator you can find, even if you have to pay a higher price for it than for a cheaply-built machine.

The next thing is the brooding. Of course, if you have a lot of broody hens, at this time, the task is a simple one. Place a few of the newly hatched chicks under each hen to be used as mother, preferably a hen that has

been sitting for a week or longer, and leave them there until the old hen, being fooled into the belief that she has hatched the lot herself, is ready to take care of them. However, this method of growing chicks is not as satisfactory as the use of a good brooder.

If the weather is yet cold and damp, at this time, I prefer to leave the hen with her brood in some out-building with a dry floor for a while, or if it is already warm and nice, the broods may go out in the open at once. An enclosed yard with coops, or an orchard, with little chick houses placed at proper intervals, offer good opportunities. Outdoor brooders located in an orchard afford an almost ideal method of growing chicks after the weather becomes settled. A home-made brooder, or cheap, make-shift contrivance is never satisfactory and is frequently dangerous. Use a first-class brooder, even if you should have to pay double the price asked for a poor one.

I am telling nothing new when I state that some old capons, if you have them, may be pressed into service as brooders. A writer in the "Country Gentleman" says: "Choose a large, fine capon, not too young. Envelop his head in your hand, and puff into his mouth and gills smoke from a tobacco pipe, the stronger the pipe the better. Shake the bird's head after each blowing, repeating for five or six times until the bird seems unconscious; then place him on the young chickens and set the box in a dark corner for some six or eight hours, or until the next morning, when this hypnotized capon will carry and care for these young birds like a hen. In my hands, this has proved eminently successful, and I commend the process to all."

The only objections I have to the plan is the great weight of the capon, and his liability to kill the little chicks by stepping on them with his whole weight. At any rate, I should prefer to use a good brooder or to let a medium sized hen take care of the brood first for at least a week after hatching, before turning the chicks over to the care of the capon. Owners of capons, however, can easily give this plan a trial, and then tell for themselves how they like it.

ARTIFICIAL INCUBATION AND BROODING.

BY THE EDITOR.

After summing it all up, there certainly is no better way for the progressive poultryman than to resort to artificial incubating and brooding, if he intends to engage in the poultry business to any great extent. In a small way, hens may answer the purpose very well, but when one gets into the hundreds of chickens, fussing with broody hens, taking care of troublesome sitters, washing the eggs, and the other many annoyances of the so-called natural method make life a burden. The modern, up-to-date incubator which requires no supplied moisture and furnishes an even, natural heat by diffusion, can be depended upon to hatch as many if not more and better chicks than the average hen. Thousands of successful incubator operators throughout the country bear testimony to this fact. It takes a little time, a little study, and a little patience to learn to run an incubator properly, but the same is true of anything else worth doing, and if it is worth doing it is worth doing well.

When the novice (or veteran for that matter) buys his first incubator, he should carefully study and follow the manufacturer's directions in every detail, neglecting no one thing no matter how trivial it may seem. He should not try experiments in his own behalf, but should operate his incubator as strictly according to the manufacturer's directions as it is possible for him to do. In this way success is certain.

He can obtain chickens when he wants them and in the number desired at the proper time, and all this is accomplished with ease and comfort, with none of the attendant dirt and distressing conditions which always accompany the practice of sitting hens. With the hen mother, you always have lice to fight; there is always a period when it is difficult to obtain good sitters; there is no known way to tell which hen is a good sitter and which is not, except by trial, and the fact that a hen has made a good mother once does not always prove that she will do so again. Again, the hen may sit quietly and well on a nest full of China eggs, and when given the real thing, in the shape of eggs which have cost you from three to five dollars a sitting, she will persist in sitting standing up, or on trampling the eggs into a custard. Even if she does not do these aggravating things she may, when the eggs come to pip, become uneasy and trample the chicks to death while they are hatching, or prove so unnatural a mother that you are

Chicks in Outdoor Brooders, Fairview Farm.

lucky if you are able to bring one or two chickens of the clutch to maturity. You have lice to fight all along the line from the time you set the hen until the chickens are ready to market.

With the incubators all these disagreeable things are avoided. You set your machine when you are ready to set it, you put in the incubator the required number of eggs, as many as from six to thirty hens could care for, according to the size of your machine. If you have learned how to operate the incubator properly, you have only to keep the lamp filled and cleaned, the regulator adjusted and turn your eggs twice a day. This means only from ten to twenty minutes work each day with your machine, including the testing, which you will do on the seventh and eighteenth days. If you have run your machine as you should and the eggs are of good quality, you know to a certainty that you will get a good hatch of strong, lively chicks that are free from lice at the close of the 21st day. There is no mussing, no fussing with a number of angry biddies, no eggs to wash, no foul nests to clean out and no worry about hens dying on the nests, sitting standing or trampling the eggs and chicks. You get your chicks when you want them, and are able to grow them to market size, so that they will be ready at the time of best prices.

This also applies to brooders. The modern brooder of best make of the three-compartment outdoor pattern will do by far better work than the average hen. It will care for fifty chicks from the shell to from eight to ten weeks of age, and care for them better than three or four hen mothers would be able to do. There will be less work for the caretaker, less waste of food, and better, healthier, stronger and sturdier chicks than are often reared by hen mothers. There is no trouble with lice, little danger of disease, and comparative ease in caring for the chicks.

Of course, it is necessary to learn to run the brooder properly, and this the operator should do, thoroughly familiarizing himself with the brooder by running it empty for three days to a week, making sure that he thoroughly understands it before he attempts to care for chicks in it. If he will do this and make sure that he has his brooder thoroughly warmed up by running it at least forty-eight hours before he intrusts chicks to it, we are certain that he will have good results. The specially prepared chick food is not devoured by a number of greedy hen mothers, but goes to benefit the growing chicks, and in place of three or four flocks to feed, the operator has only one flock of fifty chicks to the brooder. When he has made his rounds for the night and carefully put his little flocks to bed, he feels certain that he will find them all in good condition in the morning, and that the brooder can be trusted to take good care of them. With a good brooder and a good chick shelter, they can be kept safe at all times from hawks, cats, rats and other marauding animals. The brooder can be readily shifted from place to place about the farm, giving the chicks practically all the benefits of free range, while at the same time keeping them confined whenever desired. The little chicks may be given the range of the garden, the berry patch, or any other place where it may be desirable to range for bugs and worms, with no fear of a busy, hungry mother hen scratching and tearing up everything within reach. The little chicks do no harm; a few mother hens would soon uproot almost any garden.

Incubators and brooders as a means of growing chicks to make capons when the object is to market them in any considerable quantity, are an absolute necessity. They give the grower a considerable number of the chicks of the same age at the time when they are most desired, and in this way make a considerable saving in labor by having a large number of chicks ready to caponize at one time. When the day for operating comes the work can be finished up in short order, caponizing from fifty to a hundred or two of birds with very little more trouble than it would take to attend to the dozen or so odd-sized chicks grown under hens.—*Editor.*

Chick Shelter for use with Outdoor Brooder.

CHAPTER X.

DISEASES AND INSECTS AFFECTING CAPONS—PREVENTION PREFERABLE TO CURE.

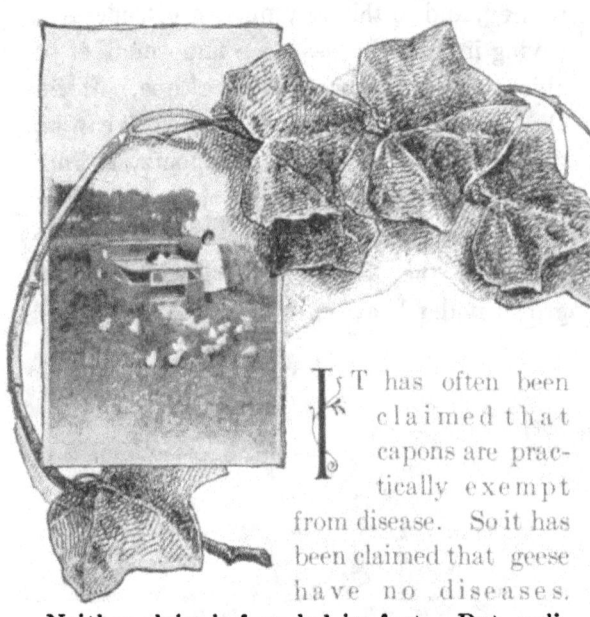

IT has often been claimed that capons are practically exempt from disease. So it has been claimed that geese have no diseases. Neither claim is founded in fact. But ordinarily you will have no more trouble with capons than with geese in this respect.

I have lost capons from disease, yet never in a larger proportion than I lost of other fowls at the same time, and never a larger percentage of the flock.

Usually my capons are quite healthy and robust. They grow fat; they get very fat, and yet are taking moderate exercise, and seem to feel well. I consider it a thankless task to "doctor" a sick fowl. The secret is to *keep* the birds in health rather than restore them to health. To dose a seriously affected fowl with medicines is usually more trouble than the bird is worth; and in many or most cases

it is difficult to properly and accurately diagnose the disease. The axe applied between head and body, and the carcass buried deeply out of sight, is generally the surest and safest cure.

With proper management in feeding and watering, with good ventilation, exercise and scrupulous cleanliness, a sick bird should be, and will be, a rare exception.

The only time that I had ever any serious trouble with sickness among my fowls, was ten or a dozen years ago when for several years I kept annually a hundred or two of Pekin ducks, running at large with my Langshans. The ducks befouled the whole place, and my splendid Langshan hens were bound to drink out of every stagnant pool and in those wet seasons, we were bound to have some such pools in the barn yards. The consequence was that I lost a large percentage of my hens by "cholera." (Probably enteritis or fowl typhoid. Ed.) Sprinkling iron and copper sulphate solution all over the premises, and spraying the mud holes with kerosene did not entirely check the trouble. It was only by getting rid of those pesky ducks that I was able to stop the epidemic, and since then I have hardly lost a bird by disease.

Cleanliness—absolute cleanliness—is the great secret of health—with proper feeding, of course. Stagnant pools are a constant source of trouble and danger. If they cannot be drained or dried up, pouring a good quantity of kerosene, or better still, a little napcreol, on the water will mitigate the evil to some extent.

The drinking fountains must be closely looked after. They need frequent thorough cleaning during the warm season, and possibly an occasional scalding, or rinsing in a solution of napcreol and water. A small piece of copperas (iron sulphate) thrown into a bucket of drinking water, occasionally, is also a good thing. Best of all devices for supplying fowls with fresh water is a running stream, a brook or spring, or a steady flow into a low dish from a hydrant that furnishes pure water.

It hardly needs special mention that the poultry houses and coops must also be kept scrupulously clean, and during the warm season well ventilated by opening the doors and windows. The gases arising from the accumulation under the perches are not particularly conducive to health. The free use of road dust or dried muck will greatly aid in keeping these accumulations dry and inoffensive. A lot of dried muck is of especial benefit as it not only keeps the droppings sweet and inoffensive, but also adds value to the manure. Land plaster is often recommended for use in stables and poultry houses. It is good, no doubt, but plain superphosphate (acid phosphate or soluable South Carolina rock) costs but twice as much, and is five times as valuable for manure. I make it a practice to clean the deposits from under the perches quite often, then scatter some road dust on the floor, and upon this a small quantity of superphosphate, repeating the application of the latter every few days at least, scattering it over the droppings. This removes all bad odors, with the exception of the bearable

one of the superphosphate, and makes a most excellent and very rich manure. Kainit may be used in place of the superphosphate, and would make a manure extremely rich in potash for soils and crops that need that particular element more than the phosphate. In using these phosphate preparations, care must be taken to use them in a finely powdered form, as, if used in small lumps they might be eaten by the fowls and would prove injurious.

In some seasons I have had a small percentage of my capons suffer from leg weakness. This may have been caused by errors in the selection of foods, or more likely, through an inherited or constitutional weakness. Sometimes the birds showed the disease only when they were almost full grown, and we have used them, or let other people use them, for the table, just the same. I have no remedy to suggest except careful selection of rugged breeding stock, and of foods and feeding methods.

In a general way I may say that insects are more troublesome in our poultry yards and houses than diseases, and we must keep them in check if we desire to have our fowls remain in health.

The gapes, caused by a worm parasite, gives a lot of trouble to people who attempt to raise chicks outdoors in some portions of the country. We do not have this enemy to contend with here (LaSalle, N. Y.) It is common

Galvanized Iron Drinking Fountains.

49

in the Southern States. The remedy is to thoroughly disinfect the ground on which the chicks run with a good fluid germ destroyer like napcreol, keep the chicks confined in small runs which have been disinfected—remove the chicks frequently to new quarters with equally restricted runs which have been newly disinfected, and thoroughly clean up and disinfect the old ground on which the chicks have been running as soon as they leave it. Perseverance in this method is the only sure road to successfully getting rid of this troublesome disease. This treatment must be continued until the chicks are well grown. Dressing the ground with air slaked lime, which is thoroughly slaked and pulverized, and plowing it in is an effective means of ridding the ground of gape worms and their eggs, but it must be repeated at frequent intervals as often as any case of gapes shows itself on the place.

Scaly leg is caused by an insect, a microscopic scab mite. It is not a dangerous disease and easily cured. The remedy usually recommended is a mixture of lard and kerosene, smeared freely on the affected parts of the bird's legs—the dose should be repeated. I always use clear kerosene. This is put into a quart or two-quart basin, filling it half full, and taken to the poultry house at night, when

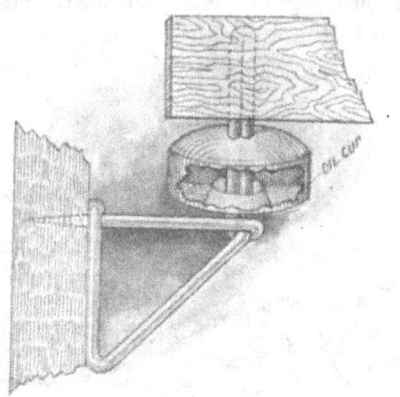

The Anti-Lice Roost Hanger.

the birds are all on the perches. One after another of them is picked off, held by the wings with one hand, while the legs are held together by the other, and the feet are then dipped down into the kerosene. Care is taken that the kerosene reaches way up to the joint. In three or four days, or a week's time, this procedure is repeated, and shortly afterwards the scales will all have disappeared and the bird's legs be clean and natural. [An ointment made of napcreol and lard applied daily, and well rubbed into the legs of affected fowls is very effective and is agreeable to use. To make the ointment, mix one teaspoonful of napcreol with a half pint of melted lard and stir the mixture until cool. Surekil lice paint mixed with kerosene in the proportion of one teaspoonful of the lice paint to a pint of kerosene makes an effective remedy for scaly legs.—Ed.]

A good lice paint is the sovereign remedy for the small mites or "red spiders" which infest the great majority of poultry houses, and do a lot of injury, sapping the vitality of our birds, and sometimes actually killing them, especially when on the nests sitting. Yet they are very easily checked, and banished from our chicken houses. In the first place, let the perches be hung on loops of heavy wire, or use the Gross anti-lice roost hangers here illustrated. Have the nests movable too. At the first sign of a mite (red spider), give to the perches and the nests and even the walls of the building, and in fact any place where a mite might find a hiding place, a thorough spraying with lice paint or with lime whitewash into which a good deal of kerosene has been mixed. Hit everything and the mites will leave. Lice paint is the thing that kills them stone-dead. If you have a small sprayer, or a compressed air sprayer or any sprayer that has a good pumping apparatus attached to it, it will take you but a little time to apply the mite-killing fluid all over the roosts and the entire interior of the room, and so do the job well. . That is the way I do it, and mites are not often allowed to suck much poultry blood in my establishment.

Hatch your eggs in incubators and bring up the chicks in brooders, away from ,older fowls, and they will not be bothered with lice. Or if you cannot do that, keep the older fowls free from lice by using a good lice paint on the roosts and fitting up a dust bath for them, and the chicks will grow up to be pullets or capons without having known a louse.

As matters stand, however, it seems to be a fact that fowls—the majority of them— have some lice, sometimes more and sometimes less. These great, big lice are very sensitive to dusty material. Dust irritates them, stops up their pores and breathing apparatus. Thus, we use dust of some kind to free our fowls of these pests. If we pro-

vide a suitable dust-bath for the birds, a big heap or box of road dust, or rather, moist earth from the garden, and if tobacco dust mixed in, all the better, the birds will pretty nearly get rid of their undesirable boarders themselves. They wallow in the dust and shake it all through their feathers, and the lice will say good-bye.

To sitting hens, however, I never fail to give an additional treatment. My favorite louse killer is a good lice powder, of which I keep a box or can always in readiness in the hatching room. From time to time I apply a small quantity to each sitting hen, by dusting it thoroughly all through the feathers of the neck, back, wings, etc., and thus kill all the lice that it reaches. The nits may escape, but will be reached by another application a few days later. In the absence of insect powder, I have used tobacco dust with satisfactory results.

51

Part of 500 Capons and Pullets, Jordan Plant, Hingham, Mass.

CHAPTER XI.

THERE IS PROFIT IN CAPONS—IMPROVED QUALITY IN DEMAND—SOME OF THE ADVANTAGES OF CAPONIZING.

BY A. F. HUNTER.

FEW poultrymen realize the great advantages of caponizing the surplus cockerels, of caponizing the males not intended to be raised for breeding purposes, and the steadily increasing demand for capons, and the steadily growing interest in the subject of caponizing for the advantage of the increased profit is good proof that the business is a profitable one. It will be a surprise to many readers to be told that there is always some sale for capons, although, generally speaking, it is the winter and spring markets that take the most of them and pay the best prices. The absence of soft roasting chickens at that time leaves a void which capons come in to fill, and there is at that time a sale to private families, as well as the regular demand for them at the hotels, restaurants, and clubs. It is not generally known, too, that some market poultry raisers are regularly caponizing all of their male birds many of which do not go to market as capons at all, they being marketed as roasting chickens of eight to fifteen pounds to the pair. The poultry raisers find that there is a decided advantage in having all of these male birds caponized, their greater docility preventing the dissipation of food energy in nagging and scrapping, and allowing of keeping large numbers together and in quite small quarters, and keeping the birds steadily growing so that they are in condition to market as soft roasters at any time that there is a good demand for them. We recently visited the poultry farm of Farrar Brothers, West Norwell, Mass., and saw some 5,000 chickens in various stages of from just out of the incubator up to chickens that would weigh seven or eight pounds when dressed. All the male birds of sufficient size had been capon-

ized, and all of the younger males would be caponized as soon as they reached caponizing age. In talking of this with the Messrs. Farrar, we learned that it was their custom to caponize all of these male birds, even although they might be marketed as roasting chickens, and they found it a decided advantage to do so. In this section of the country the bulk of the poultry kept are of the Asiatic varieties, Light Brahmas predominating, and the Messrs. Farrar find little difficulty in buying eggs for hatching which give them the desired quality from their farmer-neighbors, hence they keep little laying stock of their own. As they expressed it, they keep about 100 head of layers, Light Brahmas, chiefly for experimental purposes; they depend almost altogether upon their neighbors for the hatching eggs. They begin operations in the late summer and continue through the fall and

Part of 2,000 Half Grown Capons, Farrar Bros., West Norwell, Mass.

winter, uninterruptedly if a sufficient supply of eggs can be obtained; it is sometimes difficult in the fall to keep the incubators filled with eggs sufficiently high in fertility to justify running them. We give illustrations of some of the fine caponized birds that we saw there the last of March, which will find their way to market as soft roasters, as Messrs. Farrar find that they can get just as good prices for roasting chickens weighing eight to ten pounds each as for capons. In the long house shown in the illustration were about fifteen hundred head of half grown chickens, both cockerels and pullets, all of the cockerels having been castrated; and these birds in large flocks were living in peace and harmony together, growing into the finest quality of market poultry.

The Messrs. Farrar use a hot-water pipe brooder house for winter chickens, as they consider the hot water system more reliable and less trouble to manage than individual brooders. Last winter, however, owing to the scarcity of coal, they used the individual brooders until the coal strike was ended. As spring approaches they begin to use individual brooders in the pens of the long house shown in the picture. These pens are ten to fifteen feet in size, and 150 chicks are put in each brooder. When these chicks are old enough

Mr. A. F. Hunter and a Pair of Farrar Bros. Capons.

to wean from the brooders, the latter are taken out of the pens and the chicks left to their own devices. As, however, 150 chicks are rather too many to raise in one pen, about 50 of them are taken out and put into colony houses such as are shown in the illustration, and those 50 chickens are raised to market maturity in colony houses of six by eight feet each. This is remarkably close quarters, actually less than a square foot of floor space per bird, and yet these docile, tractable, caponized chickens living and thriving in those close quarters is good proof of what can be done if the male birds are castrated. Wishing to get a photograph of a pair of these choice chickens, two good birds were selected from the ones in this house and held in a position for the photographer to get a good view of them. Mr. Farrar said that these birds would then weigh about 9 pounds apiece, and it is a good illustration of the high average excellence of the about 50 birds in that house, that it was difficult picking out from the lot two that seemed in any way better than the others—the selection could have been made almost at random. Of course, these chickens of nine pounds weight in March, were fall hatched and were about six months old at the time of our visit. They had been grown wholly on Cyphers Chick Food and high protein beef scrap, and so handled, the Messrs. Farrar told us their losses were almost none. That fall hatched chickens can be raised to such high perfection and in very close quarters with very little or no loss is a high tribute to the management and care given them, as well as to the advantages and benefits of castrating the male birds. We have dwelt upon this story because of the light it throws upon the benefits of castrating all male birds intended for market.

Capons are in demand from December to May, with prices ranging highest in February and March,—in the latter month, capons sell at retail for from 25 to 30 cents per pound, and it is good evidence of the profitableness of market poultry raising that the Messrs. Farrar were dressing their fine capons for roasting chickens, and selling them in the market as roasters, because they could get quite as good prices and as much profit by selling them as roasting chickens as they could to sell them as capons, and that too, at the period

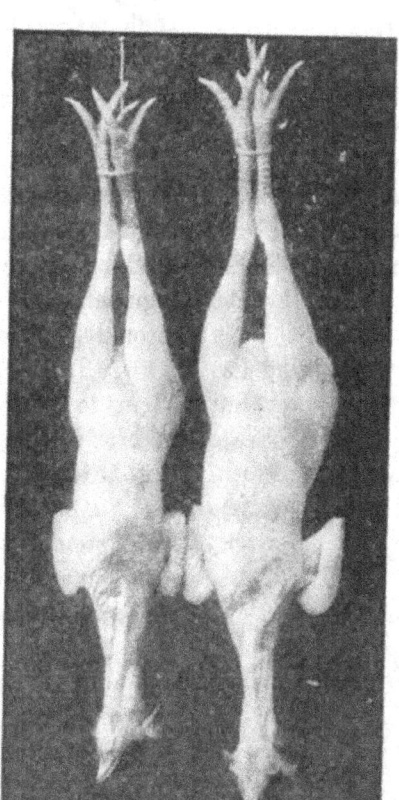

Barred Plymouth Rock Capons Dressed as Roasters.

of highest prices for capons. We believe it will be news to many poultry raisers that the Barred Plymouth Rocks will make quite as desirable capons as any chickens that are grown, good proof of which statement we present in the illustration of a pair of magnificent capons of that variety, (See page 55). These two birds were killed and dressed the day they were six months

old, they weighed alive 23 pounds, and the larger of the two weighed 11 pounds dressed. They were dressed as roasting chickens, because they were to be sent to the president and secretary of the Cyphers Incubator Company as a testimonial of regard and an evidence of the excellent results obtained in feeding Cyphers Chick Food and high protein beef scrap. Mr. Curtiss, who sent these birds to the Cyphers Company, wrote that he had handled in recent years fully half a million head of market poultry and that this pair of Plymouth Rock capons were the finest birds he had ever seen. With such a strong endorsement from a man who lives in a country where most of the birds raised are Light Brahmas, by many considered to be the very best of market poultry, we can no longer say that birds of the Asiatic varieties make the best of roasting chickens and capons. We have generally found, however, that Light Brahma or Cochin cockerels, or a first cross of these two varieties, are preferred by capon growers. In a section of South Jersey from which very large quantities of high class market poultry (known in the market as Philadelphia chickens), are shipped, we found an example of a farmer who had raised 200 capons of a cross of Light Brahma and Buff Cochin; the average weight of the 200 birds being ten pounds each, and the farmer received for them the munificent price of 23 cents per pound. This foots up $2.30 each for the 200 birds; a total of $460 by one farmer for 200 capons. When we consider that these were June hatched chickens caponized in September and marketed the last of February following, and that capons are grown at so small expense for house room, etc., it does not need an expert mathematician to discover a very substantial profit in the growing of such market poultry.

In studying the capon-market conditions in Boston, New York, Philadelphia and Chicago, we find that as a rule, the largest birds bring the best prices, that being the class of capons most preferred by high class hotels, restaurants and clubs. There is, however, a very considerable demand for a medium sized capon, weighing, say seven to nine pounds, chiefly for family trade, and this demand is a steadily growing one. It is worth noting also that it is not confined to the large cities, but is growing locally in many directions. Private families are learning to appreciate the toothsomeness and fine eating qualities of capons

Home Made Food Hopper.

and it is a not uncommon thing for the grand Sunday dinner for a family to be made from one. They are learning, too, that it is economical meat to buy, there being so little waste that a capon costing even $2.00 gives such excellent account of itself upon the table that it compares favorably with a roast of beef or leg of lamb when the cost and spending qualities are considered. This class of trade is certain to increase with the extension of the custom of caponizing, and the developing of the local trade; capons only need to be once eaten to be appreciated and the fact that it is as easily possible to buy a capon as a roast of beef or leg of lamb will create trade.

To the farmer depending chiefly upon the products of the farm for his meat supply, the argument for caponizing will appeal most strongly. We all know how dry, tough and stringy the cockerels become as soon as they begin to approach maturity, "staggy" or "hard" as the marketmen phrase it; and yet that hard, dry, tough "rooster-meat" is the best the farmer can supply his table with for many months in the year. It is so simple and easy a task to castrate those cockerels when young, and thus change them from pugnacious, nagging, bothersome roosters into tractable, docile capons, that have no other idea of life than to eat and grow, and which mature into the choicest of table meats,—it is astonishing that caponizing is not universally adopted. Whoever has wrestled with a cut of rooster-meat would certainly need to be very hungry indeed to desire another taste of it, if he realized that by a simple operation, comparatively easily performed by any one, that tough, dry meat could have been changed into an increased quantity of flesh of the tenderest and most appetizing quality. Think of what this would mean to the average farmer's family! And every farmer's family can enjoy the luxury if he will but castrate the surplus cockerels when of suitable age, which is when the combs first begin to start to develop; at that age the youngster has but very little flesh over his bones, and the operation is simple and easy and the recovery rapid. Another of the benefits of caponizing is that the hens and pullets are not continually nagged and imposed upon by the useless males. Every male bird not needed on the place for breeding purposes is useless,—and worse than useless because he dissipates the strength and vigor of the females of the flock by his impositions, and wastes their energies which should be conserved for egg-production. Every argument is on the side of castrating the surplus male birds. The quality of the meat is immeasurably improved and the quantity increased, because the capons are of a docile and tractable disposition, the food they eat goes to the making of tender, toothsome flesh instead of hard, dry, tasteless muscle. Nor is the sometime claim that the operation is a cruel one at all tenable. The birds flinch slightly at the stroke of the knife, but they will turn to eat food placed on the table within reach, even while being operated upon, proving that they feel the pangs of hunger more than the pain of the operation itself; the operation cannot be cruelly painful when they will reach for food while under the operator's knife and turn to eat and drink directly they are released from the table. Then, too, the point that castrating prevents the teasing and nagging of each other and the pullets and hens of the flock may be offset against the operation itself; certain it is that a quiet bird of even temper and disposition is the result of the simple operation of caponizing.

Dressing and Shipping.

When the capons are ready to dress for market they should be starved for from twenty-four to thirty-six hours, that the crops and gizzards may be entirely and the bowels practically empty of food. Have two strong nails about a foot apart driven into an overhead beam or girder in the picking house, (or shed or outbuilding where the picking is done), make two nooses of strong cord of sufficient length to allow the bird to hang just opposite the picker's breast, one noose for each leg of the capon. Many pickers have a piece of iron of about two pounds weight, with a hook affixed in such manner that after the bird is stuck and the legs hung in the nooses, the hook is fastened in the bird's lower bill; the weight holds the bird steadily in position for picking and makes the operation much easier.

When everything is ready, suspend the capon by the two nooses, catch hold of his head with the left hand and draw the thumb and forefinger downward until they strike the angle of the jaw, forcing the bird's mouth open, without choking it. A French killing knife with a lancet pointed blade is the best, although a sharp pen knife, with a small blade which is very sharp at the point, will do good work. The knife is first passed into the throat and with a couple of quick cuts the two large blood vessels at the side of the neck, just back of the head, are severed so that the blood flows freely. The knife is then held at an angle with the bird's bill, pointing towards the roof of the mouth in a line with the eye, and by a

Convenient Food and Water Device.

rapid movement is driven through the roof of the mouth into the brain, and a quick half-turn is given to the knife. This causes instant paralysis and renders the bird unconscious (practically kills it), a shudder passes through its frame, the muscles relax and the feathers loosen. Begin plucking at once, and if the piercing of the brain has been rightly done the feathers come easily. Avoid tearing the skin in plucking, there is little danger of this if proper care is taken. In dressing capons for market, the feathers are left on the tip joint of the wings, also on the legs for from a third to half way up the thigh, all the tail feathers including an inch or so up the back from the tail and those of the head and neck about half way down from the head. These feathers left on have a decidedly "dressy" effect and are the marks which distinctly identify the capon. Never cut the head off in killing; the head is a distinguishing feature of the bird; the undeveloped comb and wattles, which cease to grow when castration is performed, are a certain means of identification. Wash the head and mouth well with cold water, carefully removing all blood.

Do not draw the intestines if the birds are to be shipped to market, all the large markets prefer that poultry shall be undrawn; although local conditions vary somewhat, and in a local market or for a family trade it may be wise to "draw" the birds. Place the bird back down on a table or convenient box, cut carefully around the vent and draw out the intestines; these will be found covered with fat, which should be pushed back into the opening. When the end of the long intestine is reached, insert the finger and detach it from the gizzard, leaving everything else in. Usually there is a mass of fat about the opening, and if this is turned out slightly and allowed to harden in that position it gives a very rich appearance to that portion of the bird. Let the birds hang in a cool place until thoroughly cold, until absolutely all of the animal heat has left the body; when birds are packed for shipment with some animal heat still remaining, they become green and are spoiled; heedlessness in this matter has caused a loss of many dollars to amateur poultry shippers. For packing use a new box, or certainly one that is absolutely clean, and line the box with clean, white paper; never use paper that has printers' ink on it, as the ink will transfer to the skin of the birds, decidedly injuring their good appearance. Pack the birds into the box firmly, with backs up, being careful not to bruise them; the carcasses should be packed so firmly that they will not slip or bounce about in the box.

Not all of the birds castrated will be perfect capons; now and then there will be a tiny fragment of the organ remaining and this, starting to grow again, causes some little development of masculine traits and makes what is technically called a "slip." The slips should be killed and marketed (or eaten) as soon as it is manifest that they are imperfect capons; they are generally considered to be somewhat inferior to capons, but if well fattened will sell for nearly as good a price, and in some markets are quoted but one or two cents per pound lower in price.

Group of 110 Capons Sold Alive in Chicago, February, 1908 for $162.01.
Hatched in Cyphers Incubators, Raised in Cyphers Brooders,
Grown by Mrs. Jennie B. Hartman, Wellington, Ill.

Chapter XII.

MORE ABOUT CAPONS AND CAPONIZING TOOLS.

The Greiner Caponizing Set.

First in importance is the lancet-pointed knife, and the value of this shape of knife will be fully understood when it is realized that the incision in the side of the bird is always made with a drawing motion. In drawing the knife towards us we invariably lift the hilt slightly, and with the square pointed knife this causes a downward stroke of the point, which is liable to cut too deep. The lance (or round) pointed knife avoids this danger and always presents a cutting edge to the object as the knife is drawn towards the operator and hilt raised.

The next implement to use is a spring-spreader for pressing open the ribs and holding them spread apart. The spreader which we recommend to do the work has no "tips or ends" to catch a sleeve or in other ways bother the operator, while the broad and curved tips hook upon the ribs securely and hold them spread well apart.

The nippers or forceps are used to remove any particles of foreign matter (such as a feather or bit of dirt) that may fall into the bowel-cavity during the operation; it is also useful to handle a bit of dampened sponge in soaking up any clot of blood that may possibly ooze out into the abdominal cavity.

A small steel hook is necessary to tear open the thin, film-like membrane which encloses the intestines; and a curved-ended probe which is used to push back the intestines if they crowd over upon and conceal the organs; it is also used to search for any dirt or foreign matter which may by accident have fallen into the cavity.

The most important instrument of the caponizing set is the canula, with which to catch on to and draw out the testicle. This is a nickel tube about five inches long, a quarter of an inch wide at the top and tapering to less than an eighth of an inch at the bottom; it is closed at the smaller end excepting for two tiny holes at each side large enough to admit a horsehair or very fine steel wire. Years ago a stout hair from the tail of a horse made the loop part of the canula, and if one breaks, another is quickly and easily inserted and the desired loop formed at the small end of the canula. Most operators however, now use the fine steel wire in place of horse hair, and as wire is more stiff and unyielding the loop is more easily pushed over and around the organ, hence the trick of drawing it out is more easily performed.

The other requisites are a common tin wash basin, (or small tin pan), a sponge the size of one's fist and a few tiny pieces of sponge, and a few cents' worth of carbolic acid solution. A few drops of the latter added to a quart of water in a basin make it strongly antiseptic, so that it benumbs the skin and thin flesh through which the cut is made and fits it for quickly healing after the operation is performed.

Do not try to work with poor tools. No man can do first-class work or work advan-

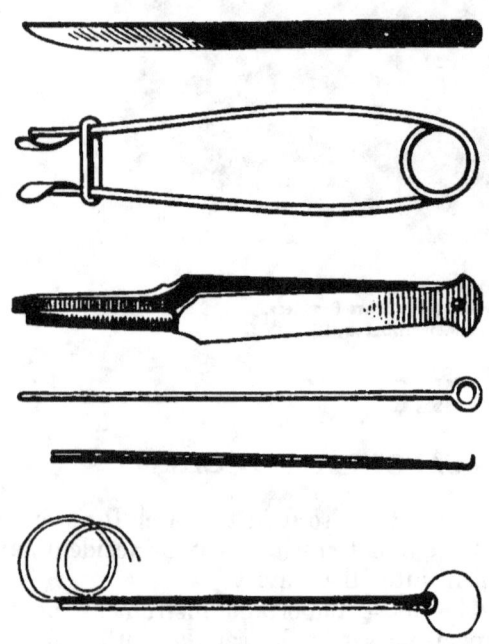

The Greiner Caponizing Set.

tageously without properly appointed tools and instruments. For the sake of humanity do not needlessly torture the birds with a dull knife or clumsy implements. When the operation is properly performed, and with a good set of tools in good condition, it involves no element of cruelty. The bird may seem to suffer very slightly when the cut is made and will possibly wince a little when the testicle is twisted off; but at other times it seems wholly unconscious that it is being operated upon, and seems more conscious of the pangs of hunger than of pain from the operator's knife.

The Cyphers Incubator Co. sells the Greiner caponizing set complete as follows:

Postpaid, in pasteboard box.........$2.75
Postpaid, packed in velvet-lined wooden
case............................. 3.00

CONCERNING CAPONS.

BY D. T. ROOTS, CONNERSVILLE, IND.
(Reliable Poultry Journal.)

Some time ago you asked me to give you our experience in the raising and fattening of capons. I take pleasure in so doing. In this article I will try to confine myself to facts and will simply relate our experience, drawing such conclusions as I think are justifiable. I wish to say that I will not try to cover up such mistakes as we have made, for mistakes are valuable assets, both to those who have made them and to those who read and profit by them.

In the early part of the summer we found on going over our cockerels that we would have something like two hundred birds that we did not consider good enough for breeders. There were a number of them that were almost good enough, but we wanted to cull closely, as it was not numbers that we wanted, but quality. The question naturally came up—What could we do with them? We could sell them in the market and get about twenty-five cents each, but this was hardly enough to make us come out even. Again, we were afraid to cull out at this time, as the birds were so young that it was impossible to tell which ones were going to make the best birds. So for the reason that they would not bring us enough money at this time and for fear of killing a good exhibition bird, we decided to see how we would come out if we caponized them, knowing that we could delay this until they were matured enough so we could form some idea as to what they would be like.

We believe that early hatched, good, strong birds weighing about two or three pounds are by far the most satisfactory size to operate on. To the raiser of exhibition and breeding stock, this is of vital importance. When the young cockerels are of that size you can make a close estimate of the ones that you should keep, and so you reduce the liability of caponizing a bird that might prove a winner. Again, birds of this size are stronger, and will more quickly recover from the shock of the operation. Of fully as much importance is the fact that in this sized birds the organs are more fully developed, and it is far easier to remove them, and not leave any part, than with the younger and smaller birds. If they are not entirely removed you will have what is known to the trade as a stag or slip, and it will bring from two to four cents less per pound.

Going back to our own trials and tribulations, we finally culled out about two hundred of our late hatched cockerels of from one and a half to three pounds each and sent for an expert to teach us how to operate on them. We found this simple, requiring mainly a steady hand and an average amount of common sense. We noticed that the larger birds stood the operation best, and that a less number died under the knife. It might be well to say that with due care the loss should not run over three per cent. Birds dying under the knife, if dressed at once, are just as good as any others, and will bring the same price.

Our next step was to purchase about four hundred more cockerels, as we wanted to find out how they would do when a large number of them were running together. To explain more fully, we all know that the liability of disease is greater in a large flock than a small one. As we wanted to learn the bad side as well as the good, we bought the above mentioned number, and after caponizing them we turned them all out in one field, the size of which was about 150 by 600 feet. In this field we placed a number of small colony houses, each 4 x 6 feet in size. We found that each house would hold about twenty-five birds, and not crowd them too much.

Barred Plymouth Rock Capon
Dressed for Market.

What the Capons Were Fed.

This field being in clover, we fed no green food. Our aim was to build up the frame and put them in good condition for the fattening process, which would begin about January 1st. We aimed to give as large a variety of food as possible, keeping in mind that the food must largely consist of what would produce bone and flesh. Our principal food was coarse oats, it being one of the foods that is high in protein and low in carbohydrates and fat. Our regular rations were about as follows:

For the morning feed, oats dry, or soaked over night. Noon only a moderate feed of wheat. The evening feed was of corn, alternated with bran mash. Plenty of grit was before them at all times, and occasionally a small amount of some poultry food or red pepper was mixed with their mash.

In a short time we found that keeping so many together would not do, for at feeding time they would trample one another, and especially the smaller ones suffered. We next built three fences across the narrow way of the field, dividing it into four parts and then sorted out the capons, putting the largest into one yard, the next size into another, etc. We at once began to see an improvement in them, and hereafter we will not place over two hundred together, and one hundred is still better. We continued feeding as above, except as the weather grew colder we increased the food, especially the corn ration. About December 15th, we removed them, placing them in two of our other buildings, which were divided into pens containing about 150 square feet each. In each of these pens, we placed twenty, which did not crowd them, especially as each pen had a yard of about five hundred square feet. At this time we began to increase the food ration, and included a feed of steamed cut clover twice each week. We also began to increase the fat forming foods by making the mash about one-half corn meal.

About January 15th, our birds were large bodied and were in fair condition, weighing from five to seven pounds, and getting better every day. From about January 21st we began to increase the corn meal until they were getting one feed per day of it. Things went on swimmingly for about one week, when they began to refuse it and some of them did not even hold their own. This necessitated a radical change in diet, and we dropped the meal for a time, giving them a variety of foods about as follows:

Morning feed, a mixture of cut clover, (steamed over night), bran and middlings. This was salted well, and twice each week red pepper was added. The noon feed was either steamed oats, or wheat. Evening feed, corn that had been boiled until it was tender. The boiled corn we found an especially good food, being easily digested, and they never seemed to tire of it. This ration we kept up until their appetites returned, then we changed our mash, adding a part of corn meal instead of the bran. I might state here that when they refused to eat, we reduced their rations so that they were always hungry.

As to the manner and amount of the regular meals, we gave them about all they would eat in the morning, a light feed at noon, and all they wanted at night. If fed in a trough, a certain amount was put in, and after a few minutes they were fed again if they seemed hungry.

Our aim has been to give them only what they could digest, believing they would gain faster than if they were surfeited or had food before them all the time. While at the Chicago show the writer was much interested in an exhibit of a concern, making a special food for fattening or rather finishing off fowls for market. We ordered a sack of it as an experiment, and finding the capons took very kindly to it, we ordered a larger quantity. We found that in a number of respects their claims were true. It produces a very fine quality of meat, and also whitens up the dark meat several shades, making a much finer appearing fowl. Birds fed with this food do not have the heavy rolls of yellow fat such as is seen on those fed largely on a corn diet, but the gain seems to be principally in the breast meat.

Our next move was to sell about 250 of them. These were not our best ones, but there were some good ones among them. Our poultry buyers had been paying from nine to ten cents for farm raised capons, but after looking over ours, offered us eleven cents, taking them as they came, large, small and slips. We next made a small shipment to Philadelphia, dressing and packing them in a careful manner. This shipment sold for eighteen cents per pound, and averaged about six and a half pounds each dressed, bringing a little over $1.00 each.

The shipment to Philadelphia was selected to be an average of our birds, and was for the purpose of ascertaining whether we could do better by shipping or by selling at home. The result was that eleven cents at the plant was equal to sixteen cents at Philadelphia or New York. This takes everything into account–loss by dressing, loss by not feeding for twenty-four hours, express, commissions, cost of picking, packing and delivering to depot. I should have stated that on February 26, we put up twenty birds and began feeding them by means of a funnel, giving each bird about one-third pint of the special food, which was mixed with water to the consistency of a thick batter.

On March 4th we dressed and shipped sixty capons to Philadelphia. This lot averaged six and a half pounds, and was made up of three lots of twenty birds each.

Pen No. 1 was fed on a mixed food, such as boiled whole corn, a mash composed of steamed cut clover hay and middlings, wheat, etc.

Pen No. 2 was fed the prepared food by means of a trough, and they were given as much as they would clean up in half an hour, morning and evening.

Pen No. 3 was fed the prepared food by means of a funnel, and they were given about a third of a pint at each feed. The results showed that pen No. 1 made the greatest gain; No. 2 next, and No. 3 least. I account for the small gain of No. 3, from the fact that handling them twice a day caused them to lose flesh. If I could have kept them longer until they had become used to the handling, they would have made a better showing. The warm weather and other causes forced us to market them before a fair trial could be given.

Now as to results. The entire sixty sold for seventeen cents a pound. I quote part of the dealer's letter in answer to my questions as to which pen was the best. "In reference to the capons, would state that there was no preference with the buyers, as to either lot, and those fed by the cheaper method sold as readily as the others. In fact, all these capons were too light in color. One great advantage in capons or any other dressed poultry for our market is to have bright yellow flesh and yellow legs. Whether this can be accomplished by a variety of food, or whether it is the breed of birds is more than I can say. The first capons that we received from you this season were nearer the requirements than this lot."

The first lot I sent were largely fed upon a corn diet, which gave the yellow color, and which seems to be the thing most desired in the Eastern market. Having eaten a bird from the first lot, and one from the pen fed on the special food, I have no hesitancy in saying that the latter bird was worth two of the former. The first one was fat and looked well, but the meat was not so tender and well-flavored as it should have been. The latter one lost very little in cooking, and a better flavored or more tender bird I never ate. A slice of the breast could be cut across the grain by simply pressing lightly on it with a fork. In other words, I mean that most of the people look at the appearance more than at the eating qualities. This will no doubt be changed as soon as the people get educated as to what a properly fattened chicken is like. At the present time, a diet largely composed of corn will produce the chickens that bring the best price.

On March 12th, we sold what remained to our poultryman, getting eleven and a quarter cents per pound. We, however, kept about one dozen that we will carry over, and will see what can be made of them by next winter. This brings me almost to the end of the chapter, but before closing I want to recapitulate a few facts, to impress them on the minds of those who have never tried this branch of the poultry industry. I say nothing to the old breeders, who have been raising and fattening capons for years, for this article was not intended for them.

Briefly, then, I would say: Caponize cockerels weighing two to three pounds, for you will have less slips and less birds will die under the knife, and they will get over the operation more quickly. Feed all they will eat up clean, during the summer of such foods as will produce a good, big frame. Do not try to get them to average over eight pounds, for the extra amount of food required will cost more than the added price per pound which you will receive. Never put over two hundred in one yard, and a less number is better. Give them all the room you can spare. To get the prices they should be kept over one season, though I doubt if the higher price will pay for the extra expense and trouble. To command the highest prices in our Eastern markets, they must have yellow skin and fat. The birds with white skin and fat do not sell so well. Allow about five weeks for fattening, and begin to ship out the best in three weeks. After they have been fed heavily about a month, it is almost impossible to make more gain and some will go backward. Keep them in a quiet place where no one will see them except the one who feeds them. The sight of a stranger will throw them back for a day or more. Be satisfied if your birds can be made to average about seven pounds. This is counting on a lot of say, five or six hundred. By careful selection of them, a higher average can be obtained. Use only well-bred stock, for you will get larger and better birds. Caponize the latter part of August, and first part of September, not when the weather is extremely hot.

We are just on the threshold of the scientific method of fattening our poultry, and while we have a work to do in learning how to properly fatten and finish them, yet it will pay us well for the trouble and time spent. An average of only one cent per pound increase would mean untold thousands of dollars in the poultryman's pocket.

TABLE OF CONTENTS.